THE BOOK THAT CHANGED AMERICA

The

BOOK
THAT
CHANGED
AMERICA

*How Darwin's
Theory of Evolution
Ignited a Nation*

Randall Fuller

VIKING

VIKING

An imprint of Penguin Random House LLC

375 Hudson Street

New York, New York 10014

penguin.com

Library of Congress Cataloging-in-Publication Data

Names: Fuller, Randall, 1963– , author.
Title: The book that changed America : how Darwin's theory of evolution
ignited a nation / Randall Fuller.
Description: New York : Viking, 2017. | Includes bibliographical references and index.
Identifiers: LCCN 2016010344 (print) | LCCN 2016046390 (ebook) |
ISBN 9780525428336 (hardcover) | ISBN 9780698186675 (ebook)
Subjects: LCSH: Darwin, Charles, 1809–1882. On the origin of species. |
Darwin, Charles, 1809–1882—Influence. | Evolution (Biology) —United
States—History—19th century.
Classification: LCC QH365.O8 F85 2017 (print) | LCC QH365.O8 (ebook) |
DDC 576.8/2—dc23
LC record available at https://lccn.loc.gov/2016010344

Printed in the United States of America
3   5   7   9   10   8   6   4   2

Set in Adobe Garamond Pro
Designed by Amy Hill

Once again, for Juliet

It was the best of times, it was the worst of times, it was the age of wisdom, it was the age of foolishness, it was the epoch of belief, it was the epoch of incredulity, it was the season of Light, it was the season of Darkness, it was the spring of hope, it was the winter of despair.

Charles Dickens, *A Tale of Two Cities* (1859)

Whenever a true theory appears, it will be its own evidence. Its test is, that it will explain all phenomena.

Ralph Waldo Emerson, *Nature* (1836)

Contents

Contents

Preface

The Civil War shattered and remade America, destroying the slave culture of the South and killing 720,000 men. It also temporarily obscured an event that was nearly as seminal: the publication of Charles Darwin's *On the Origin of Species*. Like the first mortar shell to land on Fort Sumter, Darwin's book would produce epochal change and unanticipated aftershocks. It would crash and rumble in expanding circles throughout the nation, disrupting old habits and beliefs, altering cherished ways of thinking, and remaking society. Ultimately the book would do to American intellectual life what the war did to its political, economic, and social spheres: blast it to pieces and then reconsolidate it in new ways.

One copy of the *Origin* made a disproportionately large impact on American culture. That copy—which today resides at Harvard University—was sent by its author to Asa Gray, a botanist who soon championed the new theory to general and scientific audiences throughout America. Gray passed his heavily annotated book to Charles Loring Brace, a social reformer who seized on the work as a powerful argument against slavery. Brace then introduced the same copy to three remarkable thinkers: Franklin Sanborn, a key supporter of the abolitionist John Brown; Bronson Alcott, the erstwhile philosopher and father of Louisa May; and Henry David Thoreau, who used Darwin's theory to redirect his life's work.

Today we think of Darwin's theory of evolution as the spark that ignited the battle between science and religion. But that notion overlooks the way Darwin's first American readers encountered it. Antislavery activists eagerly embraced the *Origin of Species* because they believed the book advanced the cause of abolition. By hinting that all humans were biologically related, Darwin's work seemed to refute once and for all the idea that African American slaves were a separate, inferior species. In the immediate aftermath of the John Brown affair—which all five of these early American readers of Darwin supported—evolutionary ideas were seen as powerful ammunition in the debate over slavery. Only after they had employed Darwin's theory of natural selection on behalf of abolitionism did these five thinkers come to discover that it also posed enormous threats to their other beliefs, including their faith in God and their trust that America was a country divinely chosen for the regeneration of the world.

This book is a biography of the single most important idea of the nineteenth century. It is also an account of issues and concerns that are still very much with us, including racism, one of the most intractable problems in American life, and the enduring conflict between science and religion. Ultimately, it is shaped by a notion of its own: that ideas are like the plants and animals in Darwin's theory of natural selection, shaped by their particular context, thriving or dying as a result of their ability to adapt. This process can be wrenching; it leaves people trapped between two ways of thinking and believing, stranded between two existences. But it also remakes the world.

PART I

Origins

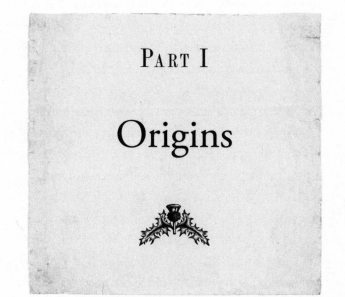

1

The Book from Across the Atlantic

The house on Sudbury Road belonged to Franklin Benjamin Sanborn, the schoolmaster of a small private academy in Concord,
Massachusetts. It was sizable, but then so was Sanborn—so tall and
lanky that a Concord farmer once referred to his legs as "God Almighty's
tongs" when he strode across the stage in a melodrama written by the
young Louisa May Alcott. At six foot four, Sanborn hulked and towered
and loomed. Yet he often felt small and insignificant, a dwarf among the
intellectual giants of a previous generation. He had been born too late to
participate in the intellectual revolution of the 1830s, when Emerson and
Thoreau and some half a dozen others had founded transcendentalism to
protest the rampant materialism of American life. To compensate for his
sense of belatedness, Sanborn rebelled. At age twenty-eight, his habitual
stance toward life was "that of opposition to whatever and whomsoever
were for the time being in the ascendant in public affairs." He flouted
tradition, courted controversy. Recently this had gotten him into trouble.
For the past two months he had lived in constant icy terror that he would
be arrested and tried for treason—a crime punishable by death. Only
through great force of will had he overcome his dread long enough to host
this evening's dinner party.

Sanborn's guest of honor was Charles Loring Brace, a member of an old, well-connected family from Connecticut. Among Brace's numerous far-flung cousins was the novelist Harriet Beecher Stowe, whose remarkable best-selling first novel, *Uncle Tom's Cabin,* had been published in 1852, one year before Brace himself embarked on the single greatest creative act of his life: founding the Children's Aid Society. Brace established a Newsboys' Lodging House to provide safe accommodations for the filthy army of homeless boys who sold newspapers throughout the city. In 1854 he began the program for which he is still remembered. Known colloquially as the "orphan trains," his Emigration Plan transported thousands of abandoned, orphaned, and runaway children to rural areas in the West, where it was believed that healthful surroundings and stable families would transform their lives.

Brace was in Concord on this New Year's Day because he had been invited by one of Sanborn's closest friends and allies, Amos Bronson Alcott, the local school superintendent, to deliver a lecture, Alcott's face resembled a rumpled pillowcase. His eyes were watery blue. He was sixty years old, and he liked to walk with a gold-headed cane that had been presented to him by a coterie of young admirers in St. Louis who believed he was the most radical idealist in America. It was true: Alcott considered ideas more real than the physical world. He believed that things—trees, apples, rivers, and swales—were little more than a wan facsimile of higher ideals. This notion went a long way toward explaining his legendary improvidence; he once embarked on a three-month lecture tour and returned with one dollar. Another time, as a young man, he spent an entire winter peddling Yankee wares to rich southern planters, sleeping in slave quarters, and borrowing books from his hosts' oak-paneled libraries, only to squander every cent he had earned on a new suit. His life would surely have been one of even greater indigence had he not married Abigail May, known as Abba and eventually immortalized as Marmee in their daughter's most famous and beloved book, *Little Women.* While Alcott conversed and dined in Sanborn's warm parlor this evening, Abba remained at home, toiling to keep their threadbare household together.

The last person at Sanborn's dinner party cared little for New York City or its burgeoning population of orphans. In *Walden, or Life in the Woods*, published five years earlier, he complained that his fellow citizens routinely asked him "what portion of my income I devoted to charitable purposes" and "how many poor children I maintained." Henry David Thoreau resented these questions. They presumed a position of moral superiority, a small-town hauteur—and they missed the central point, which was that he had found a subject far more interesting than those in need: himself. "I should not talk so much about myself," he joked, "if there were any body else whom I knew as well. Unfortunately, I am confined to this theme by the narrowness of my experience." Thoreau came to Sanborn's house that night not to learn about the Children's Aid Society or the immigrant population that made such charities necessary. He came because he had nothing better to do, because it was the first day of 1860, and because he liked to banter with Alcott and to play the wise counselor to Sanborn. He came because he was hungry.

At first the four men talked in Sanborn's parlor. Then they withdrew to the dining room. Much of what they said that evening is lost to history; much would be of little interest today. But two of the guests considered the conversation significant enough to record, and both accounts agree that eventually the discussion turned to books. Brace told of his desire to write a "Manual of Ethnology" for the general reader. Alcott mentioned an anthology of Concord authors he hoped to edit and distribute among local schoolchildren. Thoreau, the best writer of the group, kept silent about his own work. He was currently engaged in an ambitious project— a sprawling encyclopedic book about the flora and fauna of Concord that would combine science and philosophy and the pungent natural descriptions for which he is still admired—but that evening he was quiet.

At some point Brace drew everyone's attention to a book he had carried with him on the train from nearby Cambridge. He had borrowed the volume from a relative. It was thick and compact and green, printed in London by the venerable firm of John Murray, still smelling of ink, the pages cut, the spine broken, and the margins already heavily annotated in pencil. Brace placed the book on Sanborn's mahogany dinner table. The four men must have leaned forward then, their faces softening, their eyes lit with superb possibility.

The day before, a blizzard had roared out of Canada and into New England, blanketing Concord in a foot of snow. A cold snap followed. As far south as Washington, mail service was interrupted because railroad tracks were covered in hillocks of snow and ice. In Concord, along the Lexington Road and before the steep hillside cemetery that dated back to the Puritan era, spare clapboard houses emitted wood smoke from their chimneys, their windows translucent with hoarfrost. Church steeples pointed like compass needles to the sky. Abba Alcott opened her journal and attempted to sum up the year that was now rounding toward its conclusion: "The last day of the year. It seems to close with the obsequies so usual to the interment of the dead; the white shroud which nature has prepared for the remains of the old year." Snow lay piled in the fallow garden outdoors. It clung to the branches of the apple orchard and sifted onto the delicate mullions of the parlor windows. The wind grated against a corner of the house. But Abba was referring not only to the bad weather. "John Brown's martyrdom has perhaps been the event of 1859," she continued. "The hour and the man both came at last to reveal to the South their sin and to the slaves their saviour."

She was recalling the most unsettling event of that year: John Brown's attack on the federal arsenal at Harpers Ferry. Brown had long been touched by premonitions of a sacred destiny. He felt in his heart the perennial and abiding presence of the Lord, a spirit that was both all-redeeming and yet capable of inconceivable wrath. God hovered near Brown like the shimmering air above a field of summer corn, and He spoke to him through Scripture: *What is man . . . For thou has made him a little lower than the angels and hast crowned him with glory and honor!* Years ago the Lord had summoned Brown to eradicate the sin of chattel slavery, to demolish a corrupt and hellish institution that respected neither the glory nor the honor of all people as vouchsafed by the Bible. Slavery was an indelible stain upon the nation, a brutal and inhumane practice that gave lie to the United States' devotion to liberty. In obedience to his Savior, Brown had determined to attack the system of enslavement and to fulfill the promise of universal freedom laid out in the only other piece of writing he routinely read besides the Bible, the Declaration of Independence.

His plan was simple. He proposed to capture the largest federal armory in the United States and to distribute its rifles and revolvers to Virginia's slave population, which would then rise up as one to slaughter its oppressors. Brown believed that slavery was a state of war—a low, dishonest war waged under a false claim of white supremacy. Slavery was a regime of terror, a systematic attack upon the dignity of black people, a program to transform human beings into animals without souls. Despite the righteousness of his cause, Brown's raid was a fiasco. The tiny group of fighters he assembled began their assault on the armory on October 16, 1859. Failing to secure it, they retreated to a small engine house and waited for the local slave population to rescue them. They were quickly surrounded by a detachment of U.S. Marines led by Robert E. Lee, and in the skirmish that followed, Brown was injured and ten of his fighters killed. Arrested and tried, the abolitionist was hanged in Virginia on December 2, 1859.

Each of the four men who dined in Franklin Sanborn's parlor on January 1, 1860, was a radical abolitionist. All supported Brown. Sanborn had served as a member of the so-called Secret Six, a group of wealthy and well-connected men who funded the attack on Harpers Ferry. He had introduced Brown to Alcott and Thoreau a year earlier. Thoreau spoke for the rest when he called Brown a Christ returned to earth, a "Transcendentalist" who "did not value his bodily life in comparison with ideal things." Brown "did not recognize unjust human laws, but resisted them, as he was bid," having been inspired by higher promptings and an unwavering moral compass.

That night Sanborn repeated the words of a friend who believed "it wasn't Brown that was hanged but God himself." Alcott approved of this analogy because it made Brown "the Messiah of the black race, dying for their atonement." Charles Loring Brace reported that Missouri senator

Carl Schurz thought Brown's execution had "strengthened the Republicans there," a welcome bit of news in light of the coming presidential election. Then Sanborn read a letter from Theodore Parker reprinted in the latest issue of the *Liberator,* the nation's foremost abolitionist newspaper. Parker was a transcendentalist minister who had helped countless slaves escape to freedom. He was, like Sanborn, a member of the Secret Six. In his letter he predicted that slave insurrections "will continue as long as Slavery lasts, and will increase both in frequency and power, just as the people become intelligent and moral. Virginia may hang John Brown . . . but she cannot hang the HUMAN RACE."

It was a custom in nineteenth-century America to celebrate New Year's Day with lavish dinner parties. Music, dancing, flowers, and wine attended these occasions, and gifts were usually given. (The custom of giving presents at Christmastime came after the Civil War.) But there was nothing customary about Sanborn and his companions' dinner. Thoreau and Alcott were notoriously abstemious. Sanborn himself cared little for food. (He once announced at a house where he regularly dined, "I *never* eat cheese." When his hostess said, "But I've been fixing it for you for years between crackers with your beer," Sanborn replied, "Well, it's awfully good.")

Instead the men in Sanborn's parlor fed upon ideas. They dined on theories, supped on philosophy. When they had at last consumed and digested the pressing topics of John Brown, chattel slavery, and the insistent perfidy of the South, they turned their discussion to books.

Eighteen fifty-nine had been an especially good year for books. From England, Charles Dickens's *A Tale of Two Cities,* serialized weekly in the author's new literary periodical, *All the Year Round,* was fast on its way to becoming one of the best-selling works of all time. More controversial was George Eliot's first novel, *Adam Bede,* alternately praised and lambasted in the press, with one critic famously describing it as "the vile outpourings of a lewd woman's mind." (Dickens disagreed. He wrote to the pseudonymous author, saying, "'Adam Bede' has taken its place among the actual experiences and endurances of my life.") John Stuart Mill's enormously influential *On Liberty* placed a radical emphasis on the individual above the "tyranny of the majority," laying the groundwork for subsequent liberal political theory.

In America, where a nascent literary culture was beginning to assert itself with greater confidence, *The Minister's Wooing,* a sentimental romance set in Puritan New England, continued Harriet Beecher Stowe's relentless assault on the institution of slavery. Less inflammatory was Washington Irving's final volume about the nation's first president. Irving was one of America's most beloved authors, a genial and avuncular presence in the young republic of letters. Completing his huge presidential biography at the very moment the nation seemed to lurch toward disunion, he ended with an unconvincing yet hopeful assertion that the memory of George Washington "remains a national property, where all sympathies throughout our widely-extended and diversified empire meet in union." Irving died soon after this optimistic assessment.

The book Charles Loring Brace brought with him to Concord was nothing like these works. For one thing, it spoke in the comparatively new language of science, a language that could still sometimes be mistaken for literature. In the first half of the nineteenth century, science and belles lettres often cohabited; natural historians and natural philosophers (as scientists were still called) wrote for the leading literary journals while Romantic poets took scientific discoveries as fodder for their lyrics and blank verse epics. The book in Brace's possession was filled with metaphor and figurative language. Its syntax unfurled with the deliciously restrained pace of a triple-decker novel by Trollope or Gaskell. But unlike these popular narratives, its most memorable characters were plants and animals. "It may be difficult," wrote the author, "but we ought to admire the savage instinctive hatred of the queen-bee, which urges her instantly to destroy the young queens her daughters as soon as born, or to perish herself in the combat." Plants were no less a subject of wonder: "If we admire several ingenious contrivances, by which the flowers of the orchis and of many other plants are fertilized through insect agency, can we consider as equally perfect the elaboration by our fir-trees of dense clouds of pollen, in order that a few granules may be wafted by a chance breeze on to the ovules?"

Now Brace passed the book around the table, and the three other men examined it carefully. Many pages were marked in faint pencil with asterisks, checks, underlinings. Crisp exclamation marks were sprinkled throughout. The margins resembled a snowbank streaked by the prints of some flitting sparrow, and the flyleaf at the back of the book contained a catalog of questions, comments, and suggestions for future revision. Even the endpaper, oxblood in color and therefore making the penciled comments difficult to decipher, contained a pair of words positioned in the center:

Idealistic

Naturalistic

The book's front matter informed readers that the author was an English naturalist named Charles Darwin. Its ponderous title threatened to tumble beyond the margins: *On the Origin of Species by Means of Natural Selection, or the Preservation of Favored Races in the Struggle for Life.*

After dinner that night, Bronson Alcott made his way home along the frigid, snowy streets of Concord. There he performed the task he had diligently carried out for more than three decades, recording the day's events in his journal. A portion of Sanborn's dinner party, he wrote, had been spent discussing the "just published Darwin's Book on the origin of Species." He did not yet venture an opinion of the work, but his thoughts on evolutionary theory would take shape during the coming year. Alcott would never accept the book's premises. They reduced human life to chemistry, to mechanical processes, to vulgar materialism. They portrayed a world destitute of spirit. Darwin began with plants and animals, which in Alcott's opinion was precisely the wrong approach. All questions of existence were to start from above, with God and the ideal. Where, he wondered, was the soul in this new book? Where was the spark of divinity, the intuition of a higher law, the sacred principle animating the universe? Where was *transcendentalism*?

Sanborn described the book the following morning, a few hours after escorting Charles Loring Brace to the Concord train station. He sat at his desk on the second floor of his home on Sudbury Road and wrote the abolitionist minister Theodore Parker, who was in Italy at the moment. (Parker was trying to recover from tuberculosis, a disease that would kill him in several months.) Sanborn reported that a large audience had attended Brace's lecture the night before, and that afterward "Mr Alcott and Mr Thoreau dined with him here." The group had discussed John Brown and read Parker's letter, and then they conversed about "Darwin's principle of 'Natural Selection,'" which "show[s] that one race can be derived from another." Sanborn concluded by saying that Brace was especially enthusiastic about Darwin's ideas, "but you no doubt know the book."

He would never engage with the *Origin* as often or as deeply as Charles Loring Brace, who soon claimed to have read the book thirteen times. Brace would quickly incorporate Darwin's ideas into his diverse writings on charitable organizations and the study of human races. "I have been amusing myself with applying [Darwin's hypothesis] to a theory of moral and mental development of mankind," he wrote Lady Mary Lyell, the wife of the British geologist Charles Lyell. "I think it furnishes what historians and philosophers have so long sought for, a law of progress." The *Origin* confirmed Brace's belief that environment played a crucial role in the moral life of humans. It reaffirmed insights he had gained as director of the Children's Aid Society. And it seemed to offer a solution to the entrenched problems of race and slavery.

But of all those who attended Sanborn's dinner party, Henry David Thoreau would be the most powerfully affected by the new book. He was already familiar with Darwin, whose travel accounts had riveted him years earlier. In 1860 he would grapple with the *Origin*'s ideas as thoroughly and insightfully as any American of the period. By the end of January, when pirated editions of the book became available in New England, he secured a copy from the Concord Library and began taking copious notes. He worked his way through its examples and premises and began to apply them to his own research in the forests of his native Massachusetts. Five years earlier, in *Walden,* he had written: "To read well, that is, to read true books in a true spirit, is a noble exercise." Soon after Sanborn's dinner, he noted in his journal, "A man receives only what he is ready to receive. . . . We hear and apprehend only what we already half know."

Thoreau had not brought a gift to Franklin Sanborn's dinner party, but he had received one nonetheless. He had waited a very long time to receive such a book.

2

Gray's Botany

The copy of *On the Origin of Species* that Charles Loring Brace brought with him to Concord belonged to his cousin-by-marriage, Asa Gray, the Fisher Professor of Natural History at Harvard College and one of America's most prominent scientists. In November, Gray had received a letter from a colleague in England informing him that Darwin's "book is out & [has] created tremendous furore on all hands." It took another three weeks for the book to cross the wintry, storm-tossed Atlantic and to arrive at the Boston Wharf. When it at last appeared on Gray's desk, it was accompanied by a note from the author that included a plaintive request: "If ever you do read it & can screw out the time to send me . . . however short a note . . . I should be extremely grateful."

Gray was almost certainly the first American to read Darwin's *Origin* in its entirety. At age forty-nine, he was a marvel of intellectual curiosity. With no inherited wealth and few connections, he had risen from a modest Scotch-Irish background to become an endowed professor of botany at the nation's premier institution of higher learning. The poet James Russell Lowell, Gray's friend and sometime editor, once described him as "gaily innocent / And fragrant as his flowers." In truth, Gray more closely resembled a willow branch: he was flexible, yet tough. On trips to the Adirondacks or the

Swiss Alps, he often hiked thirty miles a day, lunching on cheese and apples, stopping now and then to sketch a plant or gather a specimen for his vast herbarium. His mental stamina was even more impressive. At Harvard he routinely put in sixteen-hour days, chairing countless committees, meeting with students, pledging to write more books and articles than he ever hoped to complete, tending the college's Botanical Gardens—which, under his direction, were growing beyond their available space—and spending the rest of his time hunched over his beloved brass microscope from Germany. Perpetually boyish, never weighing more than 135 pounds, he spoke too fast, his tongue tripping over his thoughts, his mind racing, his head bobbing from left to right as he tried to convey ideas as quickly as they tumbled through his mind.

Born in 1810 in upstate New York, the young Gray had found in nature an enduring source of wonder, mystery, and pleasure. He loved to ramble through the dense woods of Oneida County, exploring the foothills of the Adirondacks, collecting ferns and flowers as nearby barges stirred the waters of the newly constructed Erie Canal. In that time and place there was no such thing as formal scientific training. The word *scientist,* coined by the British polymath William Whewell in 1833, had yet to come into common usage in America. Gray worked on his father's farm, plowing the rock-strewn earth with a cantankerous horse, milking cows in a freezing shed, mending fences, and replacing the clapboard siding of the barn. But these chores bored him immeasurably, and he daydreamed about other subjects—geology, optics, astronomy: fields that fell under the vague headings of "natural history" and "natural philosophy." The first, natural history, concerned itself with *facts;* natural philosophy, on the other hand, focused on the *laws* that governed those facts. Taken together, the terms suggested that all knowledge might compose a unified whole. This idea had been eloquently expressed by the great British scientist John Herschel in his *Preliminary Discourse on the Study of Natural Philosophy* (1830), a book Gray came to know as well as the furrowed hills outside his bedroom window. Herschel explained that a "power and an Intelligence" coursed through nature, uniting it "in all departments, through which one spirit reigns and one method of enquiry applies."

If nature was immanent with divine "Intelligence," then studying it was akin to worship. Nineteenth-century scientists sought to penetrate the lush surface of nature and to discover the hidden processes of life in part because to do so was to better understand God. The passion for discovery manifested itself with particular force in America during the 1830s, when Congress funded surveys of the nation's interior and coasts, when amateur geologists, zoologists, and botanists began sharing specimens and communicating their findings through intricate networks of correspondence, and when the first American exploration of the globe commenced, returning with fifty thousand plant specimens and four thousand animals, half of which were new species. Gray was twenty-eight when the Wilkes Expedition departed. By the time it returned, four years later, Harvard had hired him on the basis of his taxonomic gifts. Gray could effortlessly summon up thousands of Latin binomial names for plants. He possessed a stunning talent for identifying and categorizing botanical specimens. He was passionate about systems, excited by patterns that revealed the fundamental orderliness of the natural world. Peering into the brass cylinder of his microscope, he searched for features that linked diverse species of sedges and mosses, studied the delicate calyxes of flowers that, under magnification, resembled jeweled scimitars and enameled wands. He was seeking similarities, shared characteristics, clues to hidden relations. With a memory that was nearly photographic, he became the preeminent classifier of plant species in North America, almost single-handedly organizing the continent's vast flora into a comprehensive system that is still considered authoritative. By 1859 he had gathered a vast treasure trove of botanical samples, classified by Linnaean principles and preserved for posterity, which formed the foundation of the work for which he was most famous, the *Manual of the Botany of the Northern United States, from New England to Wisconsin and South to Ohio and Pennsylvania Inclusive.* First published in 1848, this magisterial work was commonly referred to as *Gray's Manual.*

If Gray's mind was governed by rigor and clarity, a passionate emotional strain also influenced him. It seemed perfectly obvious to him that he dwelled in a splendid universe constructed upon some grand and still ungraspable design. This feeling drew him to the ecstatic visions of Romantic poetry, which he read all his life. It coursed even more strongly in his religious devotion. When he was twenty-four, a botanist friend converted him to Presbyterianism. Soon afterward he declared himself born again. The experience was not unusual in the "burned-over" districts of upstate New York, so called for the fiery religious fervor that had swept through the region during the Second Great Awakening and that continued to rage long after other districts had fallen into backsliding and apostasy. Before long Gray was dispatching religious tracts to his parents, in the spring of 1835 sending them Albert Barnes's lugubrious two-volume *Notes, Explanatory and Practical, on the Gospels.* He also took to signing his letters, "May the Lord prosper you and keep you all," and embraced the Nicene Creed (*We believe in one God, the Father Almighty, Maker of all things visible and invisible*). Always he carried his well-worn Bible.

Science and religion existed in a companionable relationship for Gray. Science concerned itself with the Maker's visible creation, the secondary effects of a First Cause. Religion dealt with invisible forces above or behind the natural world. Gray was not inclined to speculate on God's intentions, but he did marvel at the ingenious physical processes He employed to produce organic life. Taken together, the two ways of viewing the universe reinforced each other, endowing life with significance. By studying and organizing the botanical riches of the world, Gray believed he was exalting God's work.

He received Darwin's new book in mid-December, a week or so before Christmas. Carrying it to his cluttered study on Garden Street, in Cambridge, he placed it among the botanical manuscripts scattered across his broad table and began to read with mounting excitement. He had known about Darwin's theory for some time—had argued on its behalf several months ago at a meeting of the American Association for the Advancement of Science—but encountering the theory in print was more powerful than he had anticipated. *Here* was scientific method. *Here* was ruthless logic. *Here* were exhilarating conclusions arrived at inductively, through solid research. To make his case, Darwin had drawn from practically every branch of the living sciences, had assimilated hundreds of disparate facts, in the process creating a coherent and intellectually satisfying account of life on Earth.

As Gray immersed himself in the book, he filled its margins with exclamation marks, stars, and asterisks. He underlined sentences and marked entire paragraphs with vertical lines. Nearly every page of Gray's book is covered with the same neat, spidery penmanship with which he labeled his many botanical specimens: "Yes!" he wrote. "Well put."

The book advanced science by decades. It supplied a plausible explanation for all sorts of puzzling phenomena, opened countless avenues of inquiry. And it was beautifully written. Gray admired the work's artfully modulated tone, its modest voice, which softened the more audacious ideas rippling through the text. And there was this, as well: Darwin was a friend.

3

Beetles, Birds, Theories

When Charles Darwin was a boy, he used to ramble around the woods of his father's Shropshire estate in England in search of beetles. Years later he remembered how one day, after tearing some bark off a tree, he saw "two rare beetles and seized one in each hand; then I saw a third and new kind, which I could not bear to lose, so that I popped the one which I held in my right hand into my mouth. Alas it ejected some intensely acrid fluid, which burnt my tongue so that I was forced to spit the beetle out, which was lost, as well as the third one."

The story says a lot about Darwin—about his dogged persistence, his obsessive interests, his self-mocking humility. Born into a prosperous and somewhat eccentric family, he spent the bulk of his boyhood hunting birds, riding horses, and generally avoiding commitment toward a vocation. Such behavior provoked his father, the stern and corpulent physician Robert Darwin, to warn his second son that he was destined to be "a disgrace to yourself and all your family." Only in hindsight can we detect in the youth a glimmer of the person Darwin would eventually become: someone who adored nature and who possessed an intellect that gravitated toward questions of process, of *why* and *how*. Like some persistent leaf-eating insect, his mind bored through conceptual problems, patiently chewing over facts and evidence until a larger explanation at last presented itself.

Still, collecting beetles was not exactly a career path. With his father's encouragement, the young man went to Edinburgh, where he studied medicine and where the lectures bored him stiff and the cadavers made him ill. Again at his father's suggestion, he traveled to Cambridge to study for the ministry, hoping to become one of those rustic vicars who hastily wrote his sermons so that he might devote his time to collecting mollusks or aphids or butterflies or beetles. Around this time Darwin was visited by an enormous stroke of luck. He was invited to serve as the captain's companion aboard HMS *Beagle,* a ten-gun brig-sloop commissioned to chart the eastern coast of South America on a journey of imperial exploration and mapmaking. The expedition lasted five years and spanned the globe. Darwin, who was seasick much of the time and who would suffer from the effects of the journey for the rest of his life, nevertheless made a transformation every bit as stunning as those of the Brazilian caterpillars he gathered. He became a genius.

There are hints of this transformation in his first major book, the *Journal of Researches into the Geology and Natural History of the Countries visited during the Voyage of H.M.S. Beagle round the World, under the Command of Capt. FitzRoy, R.N.,* a work now better known as *The Voyage of the Beagle.* A gripping travelogue, the *Voyage* made its shy and reclusive author a transatlantic celebrity. Darwin observed earthquakes and volcanoes. He hiked the Andes, crossed the pampas with a band of gauchos, visited the fur-clad peoples of Tierra del Fuego, and gloried in the glaciers and the stars in the southern hemisphere. He also discovered the fossils of massive prehistoric animals, gathered geological samples from across the continent, shot and preserved thousands of brilliantly colored birds, and collected a rich cache of botanical and zoological samples for Her Majesty's museums. Never afraid to make a fool of himself while in pursuit of knowledge, he chronicled his misadventures riding the Galápagos tortoises and capturing giant lizards by their tails. And he conveyed it all in a charming, wonder-struck style.

Almost instinctively he fit his observations into larger patterns of significance. In December 1833 the *Beagle* traveled south from Patagonia. "While sailing in these latitudes on one very dark night," he recounted, "the sea presented a wonderful and most beautiful spectacle. . . . The vessel drove before her bows two billows of liquid phosphorous, and in her wake she was followed by a milky train." Most writers would have ended the account of bioluminescent creatures here, but Darwin continued to ponder the phenomenon. He noted how the phosphorescence diminished as the *Beagle* traveled farther south. "This circumstance probably has a close connexion with the scarcity of organic beings in that part of the ocean," he reasoned. Still he wasn't finished. Scooping some of the phosphorescent particles from the ocean, he observed that they "were so minute as easily to pass through fine gauze." He placed the glowing substance in a water glass, restlessly mulled the evidence. Was there a connection between the type of creatures and the colors of their phosphorescence? Was the light produced by these dead organisms part of a complicated process by which "the ocean becomes purified"? Unable to answer these questions conclusively, he finally dried the net he had used to gather the luminous matter and twelve hours later found that, when he took it up again and dipped it into the ocean, "the whole surface sparkled as brightly as when first taken out of the water."

So much of this extended passage is characteristic of Darwin: the examination of phenomena from a variety of angles, the compulsive working over of evidence until he has understood it as deeply as possible. It is typical of him that he begins at ground level—as if the tiniest, most insignificant occurrence in the physical world might enable him to unlock nature's greatest secrets. He took the same approach with the people he encountered on his voyage. In one of the book's most infamous passages, the privileged young Englishman grappled with the implications of the naked, filthy Fuegeans he encountered at the southernmost tip of South America. "I could not have believed how wide the difference between savage and civilized man," he confessed. "It is greater than between a wild and domesticated animal, in as much as in man there is a greater power of improvement." This statement, which has been understood as an insensitive and imperialistic pronouncement, quickly yields to a host of other impressions, including Darwin's admiration for the Fuegeans' ability to mimic the English language with uncanny fidelity—something no "civilized" person could accomplish half so quickly or so well.

If there is a plot nested within the *Voyage*'s pages, it is of a narrator awakening to the rich particularity of the world. That awakening led to his most extraordinary insight, famous in the annals of science, which occurred soon after his return to England, in 1836. Darwin was cataloging a number of birds he had shot in the Galápagos Islands, an isolated collection of volcanic rocks some six hundred miles off the west coast of South America, when he noticed a dozen or so finches he had collected from the various islands. Each bird was small and delicate; the biggest weighed only 38 grams. Each was dun-colored. At first glance, the birds looked identical. But upon closer inspection Darwin noticed differences in their beaks; some were larger, some sharper, some as blunt as hammers. Every island in the Galápagos apparently had a slightly different version of the finch. In an intuitive leap, Darwin seized on the conclusion that each of these different birds had adapted over time to the conditions of its particular island, producing wholly new species in the process. "Mine is a bold theory," he wrote with typical understatement in his notebook.

His name is of course forever associated with the idea of evolution (a word he seldom used, preferring instead the more accurate *descent through modification*). But the notion that a dog might derive from a wolf or that multiple species of finches might share a common ancestor—that they had *evolved* from a common source—was not particularly new, even in 1836. Decades earlier Darwin's radical and notoriously lusty grandfather, Erasmus, had postulated evolutionary development in a sprawling, whimsical, and occasionally bawdy poem entitled "The Temple of Nature, or The Origins of Society." Proposing to write an epic that would describe how organic forms "rose from elemental strife," the senior Darwin pictured the development of species as a protean, mounting process, an upward climb along "a thousand jasper steps with circling sweep / . . . winding steep."

Other eighteenth- and nineteenth-century scientists had also theorized about what was more typically referred to as "transmutation." Especially prominent was Jean-Baptiste Lamarck, a French botanist who began arguing for evolution in 1799 and codified those ideas in his 1809 *Philosophie Zoologique*. For Lamarck, the history of all living things was progressive and vertical: an upward series of transitions from simple one-celled organisms, like the amoeba or algae, to increasingly complex creatures. The most basic creatures had arisen spontaneously, but over time they had grown more intricate and specialized, developing into shrubs and trees, birds and alligators and elephants. The process that enabled organisms to change and improve was something Lamarck called the "inheritance of acquired characteristics." His most famous illustration was the giraffe. Because adult giraffes stretched their necks to browse the highest leaves of a tree, he argued, their offspring were born with longer necks. This process eventually led to the oddly shaped creatures we now recognize. Humans had arrived through a similar development, Lamarck believed, becoming the most complex and ideal species, their very existence indicating that the world was progressing from imperfection to perfection.

These ideas would be developed in greater detail in the best-selling *Vestiges of the Natural History of Creation,* published in England a decade and a half before the *Origin* and read by nearly every major intellectual on both sides of the Atlantic. Henry David Thoreau devoured the *Vestiges* soon after its publication. So did Bronson Alcott and the transcendentalist Ralph Waldo Emerson. Many read the book with a mixture of fascination and repulsion. The anonymous author of *Vestiges* presented a general theory of creation that started with nebular "Fire-mists" and continued with the appearance of successive life-forms. Most scandalously, he claimed that humans, while occupying the pinnacle of the animal kingdom and serving as "the true and unmistakable head of animated nature upon this earth," had evolved from primates. Chimpanzees and orangutans were self-evidently related to us, the author of *Vestiges* claimed, for "in our teeth, hands, and other features grounded on by naturalists as characteristic, we do not differ more from the simidae than the bats do from the lemurs."

The theories of Lamarck and the author of *Vestiges* were prompted by recent scientific discoveries that described the world as both immeasurably older than once believed and in a state of constant change. Only a generation or so earlier, geologists had begun rewriting the story of the Earth, overturning biblical accounts of creation in the Garden of Eden with descriptions of an unimaginably ancient planet. The Scottish physician James Hutton provided a glimpse into Earth's "deep time" in his 1788 *Theory of the Earth; or an Investigation of the Laws Observable in the Composition, Dissolution, and Restoration of Land upon the Globe.* His ideas were subsequently refined by one of Darwin's closest friends, the geologist Charles Lyell, whose three-volume *Principles of Geology* (1830–33) revealed a planet that was always changing. Both geologists believed that the Earth's surface constantly remade itself; huge landmasses rose during earthquakes and volcanic eruptions, only to be reduced by wind and water erosion. Continents formed and disappeared. The idea that the Earth might be millions or even billions of years old suddenly explained a host of previously mysterious phenomena. When William Buckland, an Anglican clergyman in Yorkshire, discovered the first dinosaur fossil in 1824, he initially thought he had excavated the bones of a sea monster preserved during the Flood of Noah. As Hutton's and Lyell's theories gained acceptance, however, he revised his opinion, conceding that the bones belonged to a creature far older than could be accounted for by the biblical narrative.

Darwin came of intellectual age amid these discoveries. His was the first generation of naturalists capable of surveying a landscape and imagining strange, vanished worlds filled with fantastic creatures, gaseous swamps, ferns the size of elm trees—all of which had vanished eons ago and were preserved in fossils. Veins of basalt or jagged granite suggested the upheavals of volcanoes or the grinding displacements of glaciers. Layers of sediment, mixed with the remains of tiny marine creatures, revealed the action of mud and water over impossibly long stretches of time. Suddenly the geological past had become as dramatic and compelling as a novel by Sir Walter Scott or a poem by Byron.

What distinguished Darwin's theory from its precursors wasn't his emphasis on deep time, however. It was the process he called "natural selection." This activity involved the slight, random variations observable within any species. Imagine a litter of fox pups. (The example is Darwin's.) Subtle differences will invariably exist between individuals. One pup may have longer hind legs, another a slightly thicker coat. Now imagine that these creatures prey on rabbits and the occasional hare. (Hares are faster than rabbits and, because of their coloring, blend into their surroundings slightly better.) If the rabbit population shrinks for any reason—if they are ravaged by disease, for instance—foxes with longer legs or with better eyesight will tend to succeed at chasing hares. And these foxes will survive and produce more young than their less favored siblings. Gradually, as the slower foxes die out, the population will change into one better adapted to catching hares. Given sufficient amounts of time, this population might differentiate from rabbit-catching foxes in other regions. A new variation would come into existence—perhaps even a new species.

What made Darwin's insight so radical was its reliance upon a natural mechanism to explain the development of species. An intelligent Creator was not required for natural selection to operate. Darwin's vision was of a dynamic, self-generating process of material change. That process was entirely arbitrary, governed by physical laws and chance—and not leading ineluctably, as Lamarck would have it, toward progress and perfection. The world in which natural selection functioned was a world in which change was the only constant. Natural selection was a brilliant concept, but like many brilliant concepts it assaulted long-cherished ideas and beliefs. It threatened the notion that human beings were a separate and extraordinary species, differing from every other animal on the planet. Taken to its logical conclusion, it demolished the idea that people had been created in God's image.

Asa Gray met Charles Darwin for the first time in 1839, when he sailed across the Atlantic to inspect Europe's great institutions of science. A London friend, the botanist Joseph Dalton Hooker, escorted Gray to the College of Surgeons, just off Lincoln's Inn Fields, where the two visited the Hunterian Museum, a massive collection of anatomical specimens that included the eight-foot skeleton of Charles Byrne, the so-called "Irish Giant" who had been exhibited as a freak in London back in the 1780s. Here, in one of the crowded exhibition rooms, Gray was introduced to the renowned anatomist Richard Owen, who in turn presented him to a serious beetle-browed young man described as "the naturalist who accompanied Captain King [*sic*] in the Beagle."

Both men promptly forgot each other. Gray moved on to Paris, then to Italy, where among other things he amused himself by comparing the marble sculpture of the Apollo Belvedere with Byron's description of it in *Childe Harold's Pilgrimage*. Darwin was preoccupied with yet another discovery. Several months earlier he had picked up Thomas Malthus's *An Essay on the Principle of Population*, a grim, adamantine work of political economy first published in 1798. Malthus's book contained a simple premise: "The power of population is indefinitely greater than the power in the earth to produce subsistence for man." Human reproduction outstripped agricultural food production. When this happened, a bleak struggle to exist resulted. Darwin's innovation was to extend Malthus's vision beyond political economy and into the realm of all living things. In nature, he believed, every single species competed with every adjacent species in a struggle to acquire the resources necessary for survival. A superficial glance at a grove of trees yielded a peaceful landscape: oaks, elms, plane trees scattered across a hillside. But in truth this tranquil scene seethed with competition and death. Tear off a fragment of tree bark, and you might discover beetles and centipedes and slugs, all engaged in the fierce business of staying alive long enough to produce a successive generation. Even the trees were engaged in a kind of competition. Each oak or pine over ten feet tall had survived against incredible odds, enduring frost and drought and the incessant depredations of squirrels and birds and browsing deer. Each tree had gained a perilous toehold amid the dense rootlets of grass, growing despite the shade cast by mature trees, winning out over hundreds of thousands of other seedlings and now annually producing tens of thousands of winged seeds or acorns in order to repeat the process all over again. Landscapes weren't peaceful at all, Darwin realized. They were rife with conflict and tragedy.

This observation led to another: ceaseless competition was the engine driving change and development. It had propelled the Galápagos finches along their divergent paths of transmutation, ensuring that individuals best suited to the environment would thrive and reproduce. Like a farmer sifting beans for planting, Death sifted out individuals poorly suited for survival. Malthusian theory was a crucial step in Darwin's mature concept of natural selection.

He said nothing of this to Gray in 1839. Nor did he mention it twelve years later, when the two naturalists met once again. This time they lunched with Joseph Hooker at the Royal Botanic Gardens at Kew, the celebrated herbarium created by Hooker's father, Sir William. In the interim Gray had married a Boston woman, Jane Loring, who accompanied him on this trip and recorded her impression that Darwin was a "lively, agreeable person." The English naturalist soon had occasion to recall their convivial meeting in a letter to his new American friend: "I hope that you will remember that I had the pleasure of being introduced to you at Kew," he wrote. Then he asked a favor. "As I am no Botanist, it will seem so absurd to you my asking botanical questions, that I may premise that I have for several years been collecting facts on 'Variation', & when I find that any general remark seems to hold good amongst animals, I try to test it in Plants." He then begged permission to ask Gray whether alpine plants from North America could be found in other parts of the world.

Gray, it turned out, was interested in precisely this question. In response to Darwin's query, he compiled "Statistics of the Flora of the Northern United States," which compared hundreds of plants east of the Appalachians with allied species in Europe and Japan. Darwin read the work with delight. "I have been eminently glad to see your conclusion . . . it is in strict conformity with the results I have worked out in several ways. It is of great importance to my notions." A few weeks later he hinted at his work. "Nineteen years (!) ago, it occurred to me that whilst otherwise employed on Nat. Hist., I might perhaps do good if I noted any sorts of facts bearing on the question of the origin of species; & this I have since been doing." In his next letter he enclosed an abstract that detailed the operations of something he called "natural selection," which he intended to make "the title of my Book."

This was in 1855. There would be no book, however, until the summer of 1858, when Darwin received a package that was weather-beaten and addressed in a hand he did not recognize. The small bundle had been carried across the South China Sea, passing palm-fringed atolls and green archipelagos, traveling through the monsoons of the Indian Ocean, circling the tip of South Africa, then had been carried north along the Atlantic, where it at last reached Darwin's ivy-clad home in Kent after some five months. It contained a twenty-page essay entitled "On the Tendency of Varieties to Depart Indefinitely from the Original Type," written by a tall, bespectacled explorer named Alfred Russel Wallace, who had mailed it on impulse from his remote outpost in the mountainous environs of the Malay Archipelago.

Wallace was fourteen years younger than Darwin. Poor and ambitious, his personality would forever be seasoned with a dash of resentment toward a class system that had prevented him from attending university. Born in a chill and sunless Welsh village, he had quickly cast himself off like some airborne seed, floating out into the world, collecting exotic plant and animal specimens he then sold to museums and private collectors throughout Europe. By age thirty-five, he had spent more than a third of his life in South America (mainly along the Amazon) and in the South Pacific, writing about his adventures in a series of uneven but nevertheless quite readable travelogues. Early in 1858, while collecting birds of paradise in the Malay Peninsula, Wallace was stricken by tropical fever. As Wallace sweat and shivered, wrapped in a blanket and unable to sleep, the mechanism by which plants and animals became new species suddenly became clear to him. As with Darwin, the key was Malthus; Wallace had read the political economist in his early twenties, and in his current fever it "occurred to me that [Malthusian pressures] or their equivalents are continually acting in the case of animals also." His feverish mind swarmed with visions of "enormous and constant destruction," and it occurred to him "to ask the question, why do some die and some live? And the answer was clearly, on the whole the best fitted live."

Darwin was devastated. By a nearly identical thought process, an obscure Welsh explorer had arrived at exactly the same conclusion he had developed two decades earlier. "I never saw a more striking coincidence," Darwin exclaimed soon after receiving the package. "If Wallace had my manuscript sketch . . . he could not have made a better short abstract." It took some time for this development to sink in, but when it did, Darwin despaired. "All my originality, whatever it may amount to, [is] smashed," he wrote his friend Charles Lyell. But Lyell, who was one of the first geologists to believe the world was older than three hundred million years, took a longer view. He wrote back asking whether Darwin had ever mentioned natural selection to an impartial acquaintance, someone with no vested interest in the originality of the idea, who might vouch for the priority of Darwin's claim, who might even possess written evidence of his theory.

Only then did Darwin recall the letter he had sent to his American friend at Harvard.

4

Word of Mouth

Boston in 1860 was still the intellectual capital of the United States—"the hub," as the poet Oliver Wendell Holmes famously noted, "of the solar system." New York would vie for that honor after the Civil War, when publishing and the visual arts gravitated to Manhattan, but until then Boston controlled culture. Among the largest manufacturing centers in the nation, the city produced textiles, leather goods, glass, and heavy machinery, and it transported these items around the world by ship and domestically by rail. But its most renowned export was "civilization."

Boston industry had spawned vast personal fortunes. The Cabots and Sargents acquired their wealth through shipping; the Lowells and Amorys through textiles. These fortunes in turn provided economic stimulus for all sorts of cultural investment. In 1859 construction had begun on the Public Gardens, a broad open space adjacent to the Boston Common that one day would be filled with fountains and statuary. So beautiful was this area that the poet William B. Tappan compared it to "Eden in its primal beauty." Recently installed in front of the State House was a glowering statue of the black-eyed orator, Massachusetts senator Daniel Webster, and a committee had formed to procure a massive bronze figure of George

Washington—as if to suggest that the nation's first president would have been happier farther north. Nearby at 124 Tremont Street stood the publishing firm of Ticknor & Fields, which produced handsome uniform editions by local authors Ralph Waldo Emerson, Nathaniel Hawthorne, James Russell Lowell, and Henry Wadsworth Longfellow. The city also housed two of the most influential periodicals in the country, the *Atlantic Monthly* and the *North American Review*.

There was something a bit smug and self-congratulatory about all this. In a poem about his native city, Holmes admitted that "jealous passers-by / Say Boston always held her head too high." An oft-repeated piece of doggerel made the point with greater asperity: Boston was "Where Lowells talk only to Cabots / And Cabots talk only to God." Yet despite its high-minded superiority, its chilly rectitude, and its social insularity, the city abounded in intellectual energy. A volatile mixture of transcendentalist philosophy, radical abolitionism, and religious reform characterized its culture, attracting artists, scientists, and fortune seekers from the hinterlands of the South and West.

Nowhere was that intellectual culture better represented than by the Saturday Club, a group of literary men and scientists who met informally in the second-floor dining room of the Parker House hotel on School Street. The club, notable for "having no constitution or by-laws, for making no speeches, reading no papers, observing no ceremonies," was the invention of the transcendentalist author Ralph Waldo Emerson. Once a month New England's cultural elite consumed a lavish six-course meal of goose and mutton and discussed the pressing topics of the day. Among those who regularly attended were the shaggy, abrasive, and deeply religious mathematician Benjamin Peirce; the doctor-poet Oliver Wendell Holmes; the editor and poet James Russell Lowell; the abolitionist senator Charles Sumner, who had nearly been beaten to death four years earlier by a South Carolina congressman; and Bronson Alcott. At one end of the long table sat the nation's premier poet, Henry Wadsworth Longfellow, an urbane presence in a lilac waistcoat; at the other end was Louis Agassiz, Harvard's famous zoologist, smoking a cigar and savoring his glass of sauterne.

In much the same spirit of intellectual conviviality, Asa Gray first discussed Darwin's new book on the day after Christmas in 1859. He had agreed to meet his friend and colleague Jeffries Wyman, a primate anatomist, at Harvard. Wyman was a gaunt, tubercular man with a nose as sharp as an eagle's beak. He was famous among his students for lecturing beneath the skeleton of an enormous alligator he had reputedly killed on the banks of the St. Johns River in Florida. He was even more famous throughout Boston for his expert testimony in a sensational murder trial that had riveted the community a decade earlier. The case involved two scions from the city's most prominent families: John Webster, a lecturer at Harvard's medical school, was alleged to have killed the businessman George Parkman over a debt. Before a packed courtroom, Wyman meticulously reassembled skeletal fragments taken from Webster's home, helping to prove the identity of the murder victim as well as that of the true murderer.

Wyman had invited Gray and a small group of friends to view a collection of gorilla bones and skins he had recently acquired from a French American explorer. Among those in attendance were the poet James Russell Lowell, who taught literature at Harvard and edited the *Atlantic Monthly,* and the art historian Charles Eliot Norton, who had just finished translating Dante's *Vita Nuova.* These men strolled through Boylston Hall, admiring the spacious gallery with its preserved toads and salamanders, its collection of wasps' nests, its rows of rodent crania.

According to Norton, Boylston Hall that day was as "cold & chilly as the gallery of a Roman palace in February," prompting the small group to leave the exhibition hall and to walk "down into Wyman's working-room . . . round the fire." Wyman's office smelled of formaldehyde. It was cluttered with bones and human skulls. (He also taught in Harvard's medical school, sometimes with his physician brother Morrill.) In these quarters Gray began to discuss *On the Origin of Species.* To his surprise, Norton was already familiar with the book, having just returned from England, where he had picked up a copy and skimmed it on his way back to America. Norton confessed that he admired "the patience of Mr. Darwin's research, the wide range of his knowledge & his thought." He admitted that the book would surely "help to overthrow many old and cumbrous superstitions even if it establish but few truths in their place." But he wasn't exactly thrilled by Darwin's conclusions—or rather by the conclusions barely concealed behind the scientist's careful wording. Like countless future readers who would make assumptions about the book without reading it thoroughly, he declared, "At any rate I wait to be convinced I am nothing but a modified fish."

Gray disagreed. He *had* read the book with care and knew that Darwin said nothing about the origin of human species. True, those origins were implicit. True, it was almost impossible *not* to extrapolate his theory to people. But none of that was really the point. The *Origin* brimmed with brilliant ideas, Gray believed, advancing the cause of science by decades. Speaking almost as quickly as he thought, Gray stood before the glowing embers in Wyman's grate and insisted on the work's singular importance. He explained its salient points, arguing that Darwin had resolved problems that had long puzzled scientists. He had provided *plausible* explanations for a host of issues, including the geographical distribution of plants and animals and the difficulties of classification. No doubt errors were to be found in his reasoning. Gaps remained to be filled. But this was work for future researchers. As it stood, Darwin's book represented a major accomplishment in the history of science, a significant advance in the world of thought.

Lowell was impressed by Gray's argument—or at least by the passion with which he made it. Though it would be years before the poet-editor fully accepted Darwin's theory, he now asked Gray to review the book for the *Atlantic Monthly.* Gray had already committed to writing a long essay about the book for the *American Journal of Science,* the most prestigious scientific periodical in the country, but Lowell's offer was a chance to discuss Darwin's work with a much larger audience. He agreed to do it. Darwin's book was "crammed full of most interesting matter, thoroughly digested, well expressed, close, cogent." Taken as a whole, "it makes out a better case than I had supposed possible." For these reasons he agreed to present Darwin's ideas before an American audience.

Gray departed from Wyman's office that day, planning to begin the review as soon as possible. But when he returned home to Garden Street, his wife's cousin, Charles Loring Brace, was there, young and energetic, asking, as always, if he had read any interesting new books.

Brace was thirty-three, squat, and stocky; neither his thick beard nor his receding hairline concealed the boyish enthusiasm that most people remembered him by. He had come to Cambridge to pay a visit to the Harvard Library, to call on his bright, funny cousin Jane Loring Gray, and to chat with her always-interesting husband. Asa Gray had long admired the young man's father, an amateur botanist and mineralogist; for his part, Brace was fascinated by the latest science. The two men crossed swords on occasion. Gray found his wife's cousin entirely too willing to drag new theories and discoveries beyond their intended application. "When you *unscientific people* take up a scientific principle," he reminded Brace, "you are apt to make too much of it, to push it to conclusions beyond what is warranted by the facts."

Brace stayed with his Cambridge relatives until New Year's Day, when he was scheduled to lecture some twenty miles away at the Concord Lyceum. He took Gray's copy of the *Origin* with him, reading it on the train and introducing the work to Sanborn, Alcott, and Thoreau over dinner. In a matter of days he had completely devoured Darwin's book, poring over its contents with exegetical fervor. On January 2 he returned the book to its owner and traveled on to New York.

The city felt like another country. For one thing, it was overwhelmingly Democratic. If Concord was the nominal headquarters for John Brown's most ardent sympathizers, New York's close commercial ties to the cotton industry made it far more sympathetic to the South and to slavery. As Mayor Fernando Wood observed, "The South is our best customer." Another prominent New Yorker declared, "We are neither a Northern nor a Southern city," but made it clear on which side the merchant class stood. "If ever a conflict arises between races, the people of the City of New York will stand by their brethren, the white race." New York was busier, noisier, dirtier than Boston. Unlike that tree-lined city, with its polite parlor conversations and well-meaning reform committees, Manhattan was in perpetual uproar, a clatter of carriages and trucks, a din of dockworkers and ragpickers, stevedores and fishmongers, a network of streets and avenues teeming with obstreperous shouts and the incessant hawking of wares.

Brace had moved to the city in 1848, attending Union Theological Seminary in the hopes of becoming a minister. At first he celebrated New York's democratic bustle, its colorful stir of people in wagons and carts and carriages. Walking down Broadway, he gawked "at the perfect *flood* of humanity as it sweeps along. Faces and coats of all patterns, bright eyes, whiskers, spectacles, hats, bonnets, caps, all hurrying along in the most apparently inextricable confusion." Packets and freighters filled the ports; wharves creaked with cargo. On a still day, the skies were plumed with the black smoke of ten thousand coal cookstoves. But soon he encountered the city's vast and growing population of poor and homeless. The influx of immigrants, coupled with the rapid, jolting rise of industrialization, had produced some of the nation's largest and most dangerous slums. Nearly half of New York's white population had been born on the other side of the Atlantic. Many spoke little or no English.

Instead of pursuing a respectable ministerial position back home in Connecticut, Brace embarked upon missionary work in Five Points, a section of lower Manhattan so named because of its location at the intersection of now-defunct Orange, Anthony, and Cross streets. A cramped, foul-smelling district where tens of thousands lived and died in verminous filth and violence, Five Points was a grimy nightmare of American life. In the early 1840s the novelist Lydia Maria Child visited the area and remarked on the "squalid little wretches" whose "greatest misfortune" was to *not* have been orphaned: without parents, Child believed, these children would have been placed in charitable institutions and given a fresh start in life. After his first trip to Five Points, in 1848, Brace began to see New York in a different light. The city reflected "the tremendously material side of American life," with its "endless whirl of money-getting," its commodification of people. Human misery, he realized, was the by-product of lust and greed.

He began to keep a journal. "Visited the Eleventh Ward in the afternoon," he wrote one day in the spring of 1853. "Five thousand children throng the streets at all times. The younger girls sweep the crosswalks. The older girls prostitute themselves." Picking his way through cramped thoroughfares and into lightless apartments, Brace proclaimed the miracle of God's love to anyone who would listen. He was beginning to suppose that environment determined character—that people were shaped by the privileges or abuses bestowed upon them by birthright and social position. This led him to question theological doctrines that justified such misery. And it made him determined to save the children.

In *The Dangerous Classes of New York,* his classic 1872 account of working in the slums, Brace observed, "It was clear that whatever was [to be] done . . . must be done in the source and origin of the evil—in prevention, not cure." The slums were the source of misery, the context of poverty. If children could be removed from that setting before their moral and mental faculties were permanently impaired, the suffering of Five Points might one day be alleviated. To this end, Brace founded the Children's Aid Society in 1853 to help "the destitute children of New York . . . by forming lodging houses and reading-rooms for children." The plan was to get children off the street, to provide an education that would enable them to become productive members of society. Soon he created the Newsboys' Lodging House and, a year later, instituted his Emigration Plan, which relocated abandoned, orphaned, and runaway children to rural areas throughout the expanding nation.

Returning to New York after the Christmas holiday, Brace saw its slums in a new light, thanks to the book he had borrowed from Gray. "More individuals are born than can possibly survive," Brace read in Darwin's *Origin*. "A grain in the balance will determine which individual shall live and which shall die." The passage provided a perfect gloss of Five Points. Recently the Scottish journalist Charles Mackay had visited New York and predicted that the exploding immigrant population would lead to a "struggle for existence" in the city "as fierce and bitter as in Europe." Mackay is best remembered for his book of social psychology, *Extraordinary Popular Delusions and the Madness of Crowds,* a lurid warning against the irrational impulses of mass populations. A similar anxiety motivated the article about New York's immigrant population. Mackay borrowed the phrase *struggle for existence* from Thomas Malthus, and he agreed with the political economist's conviction that poverty and starvation were the inevitable by-products of hereditary defects. The poor and miserable were poor and miserable because their forebears had been alcoholics or imbeciles or had been incapable of regulating their creaturely passions. Assistance toward such people was therefore wholly unwarranted—it was in fact a threat to society. Assistance created dependency, prolonged suffering, and enabled procreation among the poor. It encouraged immorality.

Brace thought this viewpoint absurd and cruel. He believed that all children were born possessing an innate moral sense, an intuitive grasp of right and wrong. Environmental conditions led them astray; poverty was the "grain in the balance" that destroyed a life. Every afternoon Brace left his cramped office in Astor Place and visited the city's slums, where he saw natural selection at work before his very eyes. He walked through the "bedlam of sounds" on the Lower East Side, pausing at the clamorous, huddled tenements of young Italian organ grinders, their families "talking excitedly" in rooms crowded with "monkeys, children, men and women." Leaving the Eleventh Ward, he strolled as far south as Pitt and Willett streets, just below Delancey, speaking with orphans who played among heaps of refuse and bones, their faces smeared with soot and snot. Then he continued to the First Ward, near the Battery, with its "notorious rogues' den."

In early January, soon after his return from Cambridge, he traveled north to East Forty-second Street, the location of "Dutch Hill," a camp of Irish squatters whose makeshift shanties were filled with "all the odds and ends of a great city." There he entered "a low, damp, and dark cabin," his eyes slowly adjusting to the dim room, where an old woman lay atop a bed of rags, her eyes dull, her breath ragged. Nearby stood a daughter, whose infant child had recently perished from the cold. The mother told Brace she "could hardly sleep at all for coughing, and it hurt her so here, (pointing at her breast) and she feared she was not long for this world." He sat beside her and tried to exude the balm of peace, speaking "of God and the faith in Christ, and the blessed change which would soon come to her." The words had little effect. "She answered with lack-luster eye," he wrote, telling him the priest had already administered last rites. It seemed to Brace that poverty's grim selection had determined to extinguish this branch of the human family.

The next shanty was more indigent than the last: cramped quarters empty except for a bare round table. Four children huddled around a grate where a few lumps of coal faintly glowed. "I never saw such thin pale faces," Brace wrote, "as if the hunger and want had stamped them all their lives through." He learned that their mother was dead and their father off drunk somewhere. Here again was Darwin's "grain in the balance." If Darwin's process of selection culled poorly suited creatures, then surely these children had slim chances of survival. Their parents had not thrived. Again, Brace felt the consequences of a process that, according to the *Origin,* was "daily and hourly scrutinising, throughout the world, every variation, even the slightest; rejecting that which is bad, preserving and adding up all that is good."

It was this last concept—that natural selection worked for the preservation of the good—that Brace now clung to. He looked again at the poor children. Pitiably thin and shivering in the stale, frigid room, they stood before the fire, their shadows flickering grotesquely on the wall. Were they really to be winnowed by a great impersonal force that preserved only the strongest and most adaptive creatures? Was the Almighty Creator so indifferent to the most helpless of His children? He shuddered at the notion.

Throughout the coming year, in lectures and sermons and newspaper articles, he would plead on behalf of such children, describing scenes of "houseless and frost-bitten creatures in the shanties and cellars." He insisted that the privileged, warmly dressed members of his own class could do no better than to provide a second chance for destitute children. "There is something about childish poverty that touches almost everyone," he wrote in January 1860, shortly after his visit to Dutch Hill. "We can not connect it directly with laziness, or want of foresight, or vice, and the little sufferer seems to represent to us, for the time, social evils of whose distant influence it is the innocent victim."

Darwin's *Origin* gave fresh impetus to these arguments. The book was ostensibly about plants and animals. Its author had in fact set very tight restrictions on his subject matter, scrupulously avoiding any statement about humans. Only at the very end had he allowed himself a single tentative comment related to people: "In the distant future I see open fields for far more important researches. . . . Light will be thrown on the origin of man and his history." But just as Darwin had applied Malthus's political theory to biology, Brace immediately transposed Darwin's ideas to people. Natural selection provided a gloss to the countless social problems confronting him each and every day. Like a jewel tucked away in a handkerchief, the *Origin* contained a hidden message of great importance to American life. It portrayed a world of constant struggle, of ever-present death—a world that resembled the dark, huddled shanties of Dutch Hill. But it also offered a glimmer of hope for the future.

For natural selection worked for "the preservation and adding up all that is good." Brace took this phrase to mean that human society was gradually headed toward perfection. New York's lower wards might be awash with every imaginable permutation of the foreign-born, but many of these people were adapting to a new environment, to new possibilities. The people to whom Brace ministered had left the Old World to become something altogether different. They sought transformation. They wished to become a new race: Americans.

And that wasn't even the most urgent idea he extracted from Darwin's book. The *Origin* also asserted that every living creature could trace its origin to "some one prototype." This insight confirmed Brace's own deeply held belief that humans shared a common ancestry, that the biblical account of origins, in which a single pair fashioned by God settled in an Edenic garden, was true. In practical terms this meant that every person in the world—blacks and whites and people with brown and yellow skins—was related to everyone else. For Brace, the theory of natural selection was the latest argument against chattel slavery, a scientific claim that could be used in the most important controversy of his time, a clarion call for abolition.

*O*n the *Origin of Species* swept through Boston like a choice bit of gossip. The book was discussed at dinner parties, debated in drawing rooms, argued about in salons and clubs and societies with the same ardent intensity that had been expended only a few months earlier on John Brown and his failed insurrection. Members of the city's tight-knit intellectual community parsed its contents and passed their dog-eared copies of the book from person to person. Not everyone agreed with Asa Gray that Darwin's work was a masterpiece. Various readers questioned the book's hypotheses, as well as its examples and conclusions. For some, the book was discomfiting; for others, it was downright silly. But many who waded through its four hundred pages understood that the work was something new and different, a challenge to old ways of thinking, a harbinger of the future. And many saw in its theories of competition and unending change an explanation of American life. The *Origin* did what few books ever do: alter the conversation a society is having about itself. For this reason alone it was impossible to remain neutral about Darwin's ideas.

In the study of his house on Garden Street, Gray reread the book and took copious notes. He was writing the review he had promised the *American Journal of Science,* and each time he opened the book, he felt as though he was in the presence of something important. He did not yet know that the *Origin* would mark an epoch in science, that it would alter the way humans regarded themselves and their place in the universe. Yet he felt the same electric thrill he had experienced as a young man when on his first international tour he had stood in the Vatican in Rome and gazed at the *Laocoön,* awestruck by the portrait of agony so exquisitely rendered in stone. There was something sublime about the *Origin.*

Sometime in January Gray encountered the brilliant, irascible philosopher Chauncey Wright, one of those campus fixtures who haunted the lecture rooms at Harvard. Wright was twenty-nine and already making a name for himself as a mathematical prodigy, an unconventional thinker, and a contributor to the *North American Review* and other periodicals. Gray told him about Darwin's book, and the young man quickly procured a copy. By the middle of February he had "become a convert." Ephraim Gurney, a close acquaintance of Wright's and later a Harvard dean, recalled meeting on numerous occasions in 1860 to read the *Origin* "aloud and talk . . . it over . . . interminably" with his friend. Another acquaintance recalled that for Wright, "Darwin was a thinker who fairly drew from him an unbounded homage; and this lasted till his death; I never heard him speak of anyone with such ardor of praise."

Meanwhile the primate anatomist Jeffries Wyman had borrowed a copy of the *Origin* from Charles Eliot Norton after the discussion in his office after Christmas. Norton would eventually become one of Darwin's closest American friends, visiting the scientist numerous times at Down House after the Civil War, but in 1860 he remained uncertain about the new book. The *Origin* stimulated and disturbed him; he read it dutifully but with trepidation. As soon as he passed his copy to Wyman, Harvard's scholarly community waited to learn what the famously cautious anatomist would say. By January 10, the same day a pirated American edition was published in New York, Gray wrote Darwin to report that "*Wyman*—the best of judges—& no convert, but much struck with it;—says your book is 'thundering able',—'a thoroughly scientific & philosophical work.'" (Pirated editions of British books were common in antebellum America; publishers paid no royalties for these pilfered works and could therefore supply an audience eager for the latest English books far more cheaply than they could print native authors.)

A more vocal admirer was William Barton Rogers, a geologist from Virginia who had moved to Boston in 1853 to establish a school of industrial technology. (He fulfilled his ambition in 1860, when the Massachusetts legislature contributed a large sum as well as a parcel of state-owned property near the Charles River to form the Massachusetts Institute of Technology; Rogers served as MIT's first president until 1870.) Rogers had learned of the book from scientists at home and abroad. On January 2 he wrote his brother Henry, also a prominent geologist, who was living in Scotland, "I anticipate many disciples for Darwin on this side of the Atlantic." Two weeks later Thomas Huxley—the English biologist soon to become Darwin's staunchest advocate in the United Kingdom—wrote to Rogers to say that "Darwin is the great subject just at present, and everybody is talking about it." When Rogers finally secured a copy, he immediately found himself in agreement with its basic premises. Like many other Americans, he linked Darwin's theories with the controversy over race and slavery then raging throughout the nation, describing his impressions of the *Origin* to his brother in the same paragraph where he discussed "John Brown's insurrectionary projects." Before long he had also written one of the earliest reviews of Darwin's book, for the *Boston Courier,* asserting that Darwin's views had been "presented with such fairness and simplicity" that they "will not only command the earnest attention of scientific men of whatever predilections, but will in many cases win, at least, their partial assent."

Soon the *Origin* spread beyond the small circle of Boston intellectuals, attracting readers normally uninterested in science. Young Henry Adams, twenty-one years of age and decades from becoming a renowned historian, read the book just after its American publication. In his magisterially sardonic *Education of Henry Adams,* he later described himself as a "Darwinian before the letter, a predestined follower of the tide." His understanding of the book, however, was admittedly limited. Like "nine men in ten," he felt "an instinctive belief in Evolution, but . . . no more concern in Natural than in unnatural Selection." He was mainly interested in the way Darwin's theory might be applied to his own social context, and he confessed that his attraction to evolutionary theory wasn't scientific at all. Adams was drawn to the book the way some people are drawn to burning buildings or wars: he wanted to be in the vicinity of destructive power, close to an idea capable of "wrecking the Garden of Eden."

Adams finished the *Education* in 1907, at which point he had the autobiography printed at his own expense and privately circulated to friends.

It would not become public for another decade. By that time Darwinian theory had become an indisputable aspect of American cultural life, widely accepted by scientists, sociologists, novelists, businessmen, and even mainline Protestant denominations. (The Scopes Trial, held in 1925, would soon dramatize the continued resistance to the idea by Christian fundamentalists.) But in 1860, when Henry Adams first encountered the *Origin,* this was not the case. Darwin was not yet "Darwin"; no consensus had yet formed about the validity of his book. The *Origin* was still open to a wide range of opinions and interpretations.

At the time the *Origin*'s potential for cultural destruction seemed strangely familiar to the son of a political family that had long opposed slavery. "The idea of violence was familiar to the anti-slavery leaders," he later recalled, and in 1860, the year he first read Darwin's book, the threat of aggression felt closer than ever. Increasingly the great American experiment in democracy seemed on the verge of annihilation as the nation tore itself up over slavery. The *Origin* appealed to Adams on some deeply personal level because it seemed to reflect a world in which two great peoples stood poised on the edge of an obliterating conflict. America was engaged in a struggle for its own existence in 1860. That struggle, many felt, might very well end in its extinction.

5

Making a Stir

When Franklin Benjamin Sanborn was two and a half years old, a bolt of lightning struck the chimney and rattled the windows of his family's old farmhouse in Hampton Falls, New Hampshire. His sister Sarah ran up the stairs to survey the damage and found her younger brother sitting quietly with a stick in his hand. He explained that he had made the windows shake by pounding on the floor with it. "I believed myself already capable," he later wrote, "of making some stir in the world."

Making a stir would remain Sanborn's one indisputable talent. He was a mediocre poet and a worse philosopher, but he showed unfailing genius when it came to provoking others. In later years he embroiled Concord in a bitter controversy by fertilizing his garden with his own sewage. Neighbors complained of the stench; Sanborn complained of their parochialism. Lawsuits were filed; countersuits followed. When he died at age eighty-five, his life having spanned much of the nineteenth century and just enough of the twentieth for him to recognize he didn't particularly care for it, there were few to mourn. The Concord Social Circle had unanimously voted him out of its membership, an unprecedented parliamentary move. The Massachusetts Historical Society, hoping to honor one who had contributed so much to its institution, sought to locate a sympathetic eulogist who might deliver a few favorable words at its monthly meeting—to no avail. Even Sanborn's wife had refused to speak to him for years. He compulsively burned his bridges.

It had not always been this way. As a winsome thirteen-year-old, he had encountered a book of essays by Ralph Waldo Emerson and become a transcendentalist overnight. Only one other writer—Byron—moved Sanborn as much. Emerson's *Essays* seemed to suggest the world's rich possibilities, assuring him that reliance upon one's inner principles was all that was necessary to leave a mark on history. Never a contemplative person— he felt most alive when *doing* something—Sanborn responded almost physically to the flashes of insight coursing through Emerson's works. In 1851, when he was nineteen and at Harvard, he decided to write the author. His journals were promptly filled with details of a new and worshipful friendship. "Called at Mr E-s," he wrote in 1855, confiding that the two had discussed the philosophies of Pascal and Thomas Carlyle. (One imagines Emerson did most of the talking.) Emerson considered Sanborn one of a "good crop of mystics at Harvard" and soon presented the young man to other transcendentalists in Concord. Alcott became a regular dining companion; Thoreau followed. On one occasion Sanborn wrote, "Tonight we had a call from Mr Thoreau, who came at eight and staid till ten." The author of *Walden,* he thought, "looks eminently *sagacious*—like a sort of wise wild beast."

In 1855, shortly after graduating from Harvard, Sanborn declared himself "a Theist in religion, a Transcendentalist in philosophy, an Abolitionist in politics. In a word, I am an ultra reformer on almost all points." Emerson apparently considered these sufficient qualifications to invite him to open a private school in Concord. Sanborn's students soon included the transcendentalist's children, as well as those of Henry James, Sr., Horace Mann, and in the summer of 1860, Nathaniel Hawthorne. Years later Julian Hawthorne fondly recalled that "Frank Sanborn's little schoolhouse was surrounded by the great fresh outdoors, and neighbor[ed] such abodes of felicity as the Alcotts' house to play and dance in." Louisa May Alcott, too old to attend the school when it opened, felt that during the tense and tumultuous years before the Civil War, Sanborn's school "promised the greatest interest" of anything in Concord.

It was housed in a gray building near the village center, one of the first coeducational institutions in America. Sanborn's academy offered a rigorous curriculum, room and board for out-of-town students, and a host of less conventional activities. For instance, Henry David Thoreau led students on excursions into the woods surrounding Walden Pond, where he taught them the names of plants and animals. Picnics were a regular feature. So were elaborately costumed melodramas, some of them written by Louisa May Alcott, whose younger sister May was the academy's drawing instructor. Sanborn presided over these activities with a mixture of leadership and fun. Julian Hawthorne remembered him as a "simple and conscientious master" who unfailingly displayed "manhood, fidelity, generosity and enlightenment."

Sanborn had been a schoolmaster for two and a half years when he visited the baronial estate of Gerrit Smith in upstate New York. Smith was fabulously wealthy—his father had been John Jacob Astor's business partner—and he spent much of his fortune on various philanthropic ventures, including his own ill-fated presidential campaigns. He ran repeatedly on the Liberty ticket, advocating universal suffrage and the immediate abolition of slavery. In 1846 he donated 120,000 acres in upstate New York to poor blacks in the hope of creating a self-sufficient community. It was Smith's belief that if given the chance, African Americans could control their destiny and at the same time contribute to the larger economy. The experiment failed (like many idealists, Smith had not worked out the details, including the poor soil of the Adirondacks), but he continued to support the cause of abolition. To this end, he had invited Sanborn and several others to meet an extraordinary man who wanted to destroy slavery.

Sanborn had already volunteered for the Massachusetts State Kansas Committee, where he helped funnel financial aid and contraband weapons to freedom fighters determined to keep slavery out of that battleground state. Now Smith introduced him to someone who had actually fought in Kansas, a rawboned man in a battered corduroy suit. John Brown fixed Sanborn with "piercing gray eyes" and shook his hand. Then, without indulging in small talk, he announced that he needed money to purchase rifles, pistols, and cutlasses. Sanborn was impressed. "He was, in truth," he later recalled, "a Calvinistic Puritan, born a century or two after the fashion had changed." Sanborn might not have known about the incidents at Pottawatomie Creek, where Brown and his men hacked five men to death with broadswords, but if he did, he rationalized the violence. Brown, he wrote, "saw with unusual clearness the mischievous relation to republican institutions of negro slavery, and made up his fixed mind that it must be abolished."

John Brown had already drawn up a constitution for the government he planned to create in the newly freed South. (He read it aloud to Smith and Sanborn.) He described his plan to instigate a slave rebellion and asked for eight hundred dollars. Sanborn tried to poke holes in the scheme. It was too expensive, he argued. Even if enough weapons could be purchased, there was no guarantee that a slave population would risk certain execution for a white man's crusade. But Brown swatted these objections aside.

At some point Smith and Sanborn excused themselves to discuss the matter alone. They agreed the venture had little chance of success. They also admitted they respected Brown. "You see how it is," Smith told Sanborn; "our dear friend has made up his mind to this course, and cannot be turned from it." Together they helped form the Secret Six, a group of wealthy and well-connected radical abolitionists who agreed to supply Brown's small army with rifles, pikes, and matériel. The other members of the group included Thomas Wentworth Higginson, the popular nature writer for the *Atlantic;* Theodore Parker, a firebrand minister who led one of the largest congregations in the country; Samuel Gridley Howe, a physician in Boston and the husband of Julia Ward Howe, who would soon write the popular "Battle Hymn of the Republic"; and George Luther Stearns, an industrialist who had made a fortune fabricating lead pipes.

For the next year and a half the Secret Six funneled money to Brown without knowing precisely when he intended to strike. In June 1859 Sanborn believed that the insurrectionist "means to be on the ground as soon as he can, perhaps so as to begin by the 4th of July." But the attack was postponed. Then on a bright sunny Tuesday in October 1859, as he sat in his schoolroom making plans for the "annual chestnutting excursion," he learned that Brown had been captured at Harpers Ferry. Details of the failed raid soon appeared in newspapers, which reported that a trunk of Brown's correspondence had been seized as evidence from the farmhouse he used as headquarters. Sanborn's letters were among them. That night the schoolmaster stayed up late burning all papers that "might compromise other persons," and later in the week, as he sent his pupils and teachers on the annual picnic, he hired a chaise and traveled to Boston, where he consulted with a lawyer, John A. Andrew, who offered a sobering assessment of his situation. Lending material support to Brown made Sanborn guilty of treason—a crime punishable by death. He "might be suddenly arrested and hurried out of the protection of Massachusetts law."

Sanborn bolted to Quebec. "The whole matter was so uncertain," he later explained, "and the action to be taken by the national authorities, and by the mass of the people was so much in the dark, that it was impossible to say what might be the best course." Apparently he had overreacted; no federal marshals arrived in Concord to arrest him. Within a week he was back in town, where he tried to put on a brave front. "If summoned as a witness I shall refuse to obey," he wrote Theodore Parker that winter. He took to carrying a pistol, boasting that he would tell any arresting officer "he does it at his peril for I will certainly shoot him if I can. . . . I shall resist to the uttermost, and probably kill or wound my captor!"

This was his frame of mind on New Year's Day, when he hosted Charles Loring Brace.

In later years Sanborn routinely tossed out Darwin's name as shorthand for a materialist philosophy that valued physical processes over spiritual truths. But in 1860 he was too rattled by the Brown affair to fully appreciate the significance of *On the Origin of Species*. If he did not grasp the theory in its entirety, he did understand that Darwin's book seemed to describe the world he inhabited. The depiction of constant struggle and endless competition in the *Origin of Species* perfectly captured what it felt like to live in America in 1860.

Sectional tensions over slavery had simmered for decades. Sanborn had grown up during a period when North and South viewed each other with mutual suspicion, even hostility. "The Carolinian is widely different from the Yankee," declared the *New-England Magazine;* the New Orleans–based *DeBow's Review* mockingly suggested how easy it was to tell "the genuine Yankee from the rest of his species as if he were an *oran-outang,* or a South Sea Islander." America was split between two distinct cultures, two separate and utterly incompatible types of people. As the *New-York Tribune* put it, "We are not one people. We are two peoples. We are a people for Freedom and a people for Slavery." The *Charleston Mercury* concurred: "On the subject of slavery, the North and South . . . are not only two Peoples, but they are rival, hostile Peoples."

Darwinian theory suggested that competition was always most fierce between two closely allied species. The organism best adapted to its circumstances invariably won out over competitors—even competitors that were nearly identical. Sanborn seems to have intuitively seized upon this element of Darwin's argument and to have translated it in terms of the national debate over slavery's expansion and continued existence. Shortly after his New Year's dinner party, he wrote a friend, the geologist Benjamin Smith Lyman, informing him that "Mr Brace brought a book here of Darwin the English botanist [*sic*], advocating the principle of 'Natural Selection,' as he calls it." That principle fascinated Sanborn because it insisted that survival and progress were fueled by endless combat, by struggles to the death. Applied to the ongoing contest between pro- and antislavery forces, Darwin's theory could be used to justify Sanborn's participation in funding the attack on Harpers Ferry. After all, he and the rest of the Secret Six believed that John Brown was pushing the nation closer to the "irrepressible conflict" predicted a decade earlier by New York senator

William E. Seward. By encouraging racial conflict, Brown and his supporters hoped to complete the revolution begun almost ninety years earlier
with the battles of Concord and Lexington.

If Sanborn considered Brown's attack a salvo in the struggle for national
survival, many others believed it signaled an America on the verge of
extinction. The poet Walt Whitman later described the period between
John Brown's arrest and the start of the Civil War as a "year of meteors!
brooding year!" It was a time "all mottled with evil and good," a "year of
forebodings!" marked by partisanship and anxiety for the future of the
world's only democracy. In Whitman's poem about this uneasy period, the
central image is a meteor shower, a harbinger of the coming war. John
Brown is presented as "cool and indifferent" on the scaffold, certain that
his efforts to foment slave revolt are righteous and will be vindicated in
heaven. Juxtaposed against Brown's firm conviction are the "comets and
meteors transient and strange" that streak overhead. These "balls of
unearthly light" may portend the end of slavery or the end of the nation—
they resist easy interpretation—but they certainly resemble the cannon
fire that would soon produce the unspeakable slaughter of the Civil War.

Interest in Harpers Ferry did not end with Brown's execution on December 2. The insurrection remained front-page news well into the new
year as a congressional committee headed by John Mason of Virginia sub-
poenaed witnesses suspected of conspiring with Brown. Thomas Wentworth Higginson was convinced that "no one who stands his ground will
be molested . . . Mason does not wish to have John Brown heartily
defended before the committee & the country—nor does he wish to cause
[a riot], either in Massachusetts or Washington. He wishes simply to say
that he tried for evidence & it was refused him. If his witnesses go to
Canada or Europe, he is freed from all responsibility." It was Higginson's
distinct impression that to flee the country before being summoned was
the equivalent of declaring one's guilt. Moreover, it was cowardly. "Sanborn," he scolded the young schoolteacher after he dashed off to Montreal
in late October, "is there no such thing as *Honor* among confederates?"

But on January 16 Sanborn's worst fears were realized. Walking to the post office in Moses Prichard's general store in Concord, he heard a strange voice call his name. Sanborn turned and said, "How do you do?" In response the stranger handed him a summons. The slip of paper announced that his appearance before Senator Mason's committee was scheduled for eight days hence, on January 24. Sanborn wrote back to Mason, proposing to testify in Massachusetts, "through fear of lack of protection in Washington." Mason assured him he would personally vouch for the young man's safety. "I was not so much concerned for that," Sanborn sniffed, "as resolved never to testify before slaveholders in regard to my friends."

When the Senate voted a few days later to arrest Sanborn, he left town once more, heading to Canada for a second time and "choosing [not] to be seized," as he put it, "before I was quite ready." He tried to justify his actions. Abolition was a holy cause, a crusade for justice. To fight for that cause—even to die—was noble. But to fight was necessary. There were those who felt that Brown's actions at Harpers Ferry were grossly immoral. Did slavery's wrongs justify the deaths of soldiers and innocent bystanders? Was the widespread violence Brown wished to provoke defensible? Sanborn believed the answer to these questions was *yes*. After all, pro-slavery fighters in Kansas had used violence to perpetuate a system that was itself founded upon the most extreme forms of brutality. How many mutilations and murders had been perpetrated by slave owners in the past two hundred years? How many women raped, how many children sold from their families? Politics had failed to solve the crisis, as had efforts to abolish slavery through moral suasion. But if fighting was necessary, what was Sanborn doing in Canada? The struggle to survive surely couldn't be won in retreat. Sanborn argued that he was saving himself for the greater fight that seemed just on the horizon. As he told Higginson somewhat defensively, "there are a thousand better ways of spending a year in warfare against slavery than by being in a Washington prison."

6

A Night at the Lyceum

On January 22—the same day Sanborn fled to Canada for the second time—Amos Bronson Alcott walked into town, ascended the steps of the Concord Lyceum, and took a seat near the podium. He clasped his gold-headed cane before him, nodded his head to neighbors and acquaintances. Whenever the opportunity presented itself, he talked.

A contemporary once said of Alcott that he was the "most adroit soliloquizer" he had ever met. He added that his conversations were in fact "versations" without the requisite "con." The reference was to Alcott's monthly "Conversations," for which subscribers paid several dollars a season to hear him discourse on subjects ranging from "The Old Testament" and "Impersonality" to "The Origin of Evil" and "The Life of Christ." "Talking," Alcott liked to say at the commencement of these occasions, "is the mightiest instrument which the soul can wield."

Some felt he wielded the instrument a bit too readily. Nathaniel Hawthorne, who moved into the old farmhouse next to the Alcott home in the summer of 1860, detested his neighbor's incessant chatter. On summer mornings, as he paced the hilltop behind his house, he turned in the other direction whenever the older man approached—"disappearing like a hare into the bush when surprised," Alcott observed without the least bit of self-consciousness. Years later, after meeting a surprisingly garrulous Emily Dickinson in person, Thomas Wentworth Higginson informed his wife that no one had "drained my nerve power so much" with talk—with the possible exception of Bronson Alcott. Others, like Emerson, believed his talk was at least better than his writing. The *Boston Post* once referred to Alcott's collected essays as "a train of fifteen railroad cars with one passenger."

He had been born in the last month of the eighteenth century on a hardscrabble farm in Connecticut. So meager and thrifty was the household that the young boy practiced his alphabet with a stick in the snow or with scraps of chalk on the floor of the family home. At eighteen he left New England to become a Yankee peddler, a common get-rich scheme in a region where the soil was too poor to farm beyond subsistence levels. Alcott sold pins, needles, combs, scissors, buttons, thimbles, thread, clocks, bobbins of yarn, tin mugs and plates, knives, razors, paste jewelry, and other housewares produced in New England. After several years of traveling through the Tidewater region with nothing to show for it, he turned his attention to the profession for which he would become most famous: teaching.

He took his inspiration from the Swiss philosopher Johann Heinrich Pestalozzi. In an era when most children sat on hard benches and learned by rote memorization, Pestalozzi believed that students should be allowed to take breaks and to sit comfortably. He thought a child's mental apparatus was more successfully developed by Socratic conversation than by mindless repetition. Above all, he argued that kindness—not the prevailing corporal punishment—was the best incentive for learning. Alcott applied these principles to a series of schools that were revolutionary for their time and, as a result, invariably closed once parents got wind of his unconventional approaches. (One student later called him "the most eccentric man who ever took on himself to train and form the youthful mind.")

In 1834, the same year he met Emerson, Alcott opened the famous Temple School in Boston, a small academy that drew thirty students from the city's wealthier families. He filled his school with busts of Plato, Shakespeare, and Jesus. When his pupils asked questions, he refused to answer directly, believing he could elicit the truth from them by asking what *they* thought. When students disobeyed, he called them to the front of the class, handed them a ruler, and told them to administer a blow to his hands. (Alcott thought that inflicting pain on someone else was a far worse punishment than receiving pain.) All these tactics were guided by Alcott's transcendentalist conviction that each pupil possessed a spark of divinity that needed gentle cultivation. "Every soul feels at times the possibility of becoming a God," he wrote, "she cannot rest in the human, she aspires after the Godlike." True education, he thought, fanned into flame the divine spark flickering within everyone.

He was the most radical transcendentalist in America. Transcendentalism had emerged in the mid-1830s as an intoxicating set of philosophical, literary, and spiritual tendencies unified by discontent with American life. The movement included people who don't usually belong to movements: mystic visionaries, hermits, and blue-sky utopianists. But it also included radical abolitionists, advocates of free love and pacifism, educational reformers, and women's rights activists. Most of the original members were affiliated with the Unitarian Church in some way or other, and these ministers, writers, and philosophers met in one another's homes, attended sermons and lyceum lectures, and filled newspapers and journals with examples of the "New Thought."

Chief among the loose coalition of intellectuals was Ralph Waldo Emerson. Tall and narrow-shouldered, charismatic and brilliantly insightful, Emerson presided over American culture like some alabaster statue clad incongruously in a black frock coat. His first book, a slim volume entitled *Nature,* was published in 1836, the same year Charles Darwin returned from his voyage aboard the *Beagle.* Revolutionary in its aspirations, *Nature* served as transcendentalism's founding document, its origin myth. It began with a simple passage designed to arouse American readers from their complacency: "Our age is retrospective. It builds the sepulchers of the fathers. . . . Why should not we also enjoy an original relation to the universe?"

Emerson believed, like Alcott, that God was present everywhere, especially in each human soul. But divinity could be perceived through intuition and inspiration only, not through reason, and it was best found in nature, where God resided in His most unmediated form. Transcendentalism owed a debt to German idealists such as Immanuel Kant and Johann Gottlieb Fichte, who rejected John Locke's claim that environment shaped the human mind and argued instead that the mind was resplendent with powers and insights wholly distinct from the external world. Like William Wordsworth, who claimed that infants came into the world "trailing clouds of glory," Emerson believed that much of maturation was loss, that we spend far too much of our adult life in a mundane world filled with dry and dreary quotidian. What watered these barren seasons of the soul was the occasional mystical experience: an ecstatic, light-filled state in which one felt catapulted, in his words, to "the top of our beings," so that "we are pervaded, yea, dissolved, by the Mind." In one of the most memorable passages in *Nature,* Emerson transformed an ordinary walk in November into a visionary event, a rapturous experience in which he seemed to become a "transparent eyeball" while "the currents of the Universal Being circulate through me." It was an odd image—one soon burlesqued by the cartoonist Christopher Cranch—but it was meant to suggest a cosmic unity, the merging of self with a larger spirit, the "wild delight" that befell a person surrounded by a natural world suffused with divinity.

Throughout the 1830s and '40s countless young men and women were drawn to Emerson's message of spiritual freedom and social nonconformity, and his white frame house in Concord quickly became their headquarters. But the movement also met with fierce resistance from conservative thinkers. Francis Bowen, Harvard's professor of religion and moral philosophy, complained that Emerson and his followers rejected "the aid of observation, and will not trust to experiment. The Baconian mode of discovery is regarded as obsolete." Bowen was defending the methods of science against a movement that claimed that true knowledge came from within. He became especially outraged when Emerson, speaking by invitation at Harvard's Divinity School, boldly asserted that God resided within every human being and that intuition was a better guide than religious doctrine. Belief in the miracles performed by Jesus was unnecessary, Emerson argued; the moral life of the individual was miracle enough. Bowen's colleague, the Unitarian minister Andrews Norton, declared Emerson's address "the latest form of infidelity." Emerson was banned from Harvard for the next thirty years.

Bronson Alcott went further even than Emerson, considering material reality at best a nuisance, at worst an illusion. Alcott believed the universe was nothing more than a symbol of divine thought. His students perceived this symbolic world because their minds had been designed to do so. Alcott accordingly instructed them to be on the lookout for "a fuller *Revelation* of the *Divinity*!" "As an acorn reminds you of an oak," he explained, "so does the spirit within remind you of God." As his pedagogical theories developed, he paid less attention to principles of mathematics and grammar and strove instead to instill moral perfection in his pupils. "Our thoughts are the offspring of that divine power," he wrote in 1830, "which, when freed from the obstacles of human authority, and the influences of human circumstances, feels itself the agent of its own advancement, and the source of truth and virtue."

One of Alcott's assistants, Elizabeth Peabody, published an account of the school in 1835. Peabody was one of three remarkable sisters: Mary, a teacher and author who married the educator Horace Mann, and Sophia, a painter, who married the famous novelist Nathaniel Hawthorne. Elizabeth's book of Alcott's pedagogical methods, *Record of a School: Exemplifying the General Principles of Spiritual Culture,* was met with interest by the transcendentalists and with outrage by practically everyone else. Visiting the Temple School on her trip through America, the British author Harriet Martineau sniffed that Alcott "presupposes his little pupils possessed of all truth; and that his business is to bring it out into expressions." A Boston lawyer was less charitable; he bought 750 copies of the *Record* to use as toilet paper. Particularly scandalous was Alcott's discussion of sexual reproduction, which prompted a wholesale exodus from the Temple School. When Alcott tried to counter this trend by enrolling an African American girl, he lost the rest of his students. Within weeks his teaching career was over.

Now he sat in the Concord Lyceum, waiting to hear a lecture by Thomas Wentworth Higginson, the former minister and one of the most popular authors at the *Atlantic Monthly*. Higginson regularly contributed essays on physical fitness, women's rights, slave revolts, and the flowers and birds of his native Worcester. He was an indefatigable abolitionist. Throughout the 1850s he wrote countless editorials for New England newspapers, proclaiming the evils of slavery and predicting that one day the nation would become a utopia of freedom for white and black people. He had visited Kansas, supplying antislavery settlers with knives, revolvers, and Sharps rifles in their guerrilla war with pro-slavery forces, and had led an assault against the Boston Court House in an effort to free an escaped slave who had been arrested under the Fugitive Slave Act.

On this evening the Lyceum was filled to capacity. After all, Higginson had been closely involved in the Brown affair. He was a hypnotic speaker. Tall, handsome, and indifferent to danger, of all of Brown's conspirators he alone had refused to leave the country or retreat from the public eye. The Lyceum was packed with people who thought he would surely speak about the recent events at Harpers Ferry. Instead Higginson lectured on "Barbarism and Civilization." He began with a reference lifted directly from Charles Darwin's *Voyage of the Beagle*. A "certain race of wild creatures" had been discovered in Tierra del Fuego, he informed his audience. These creatures were "dark, wrinkled, and hairy." They slept in trees and caves, survived on snakes and vermin. They "cannot be tamed, nor forced to any labor; and they are hunted and shot among the trees, like the great gorillas, of which they are a stunted copy." Surprisingly, they were human beings.

Higginson may have remembered Thoreau's comment about these people in *Walden* (he admired Thoreau above all other writers): "Darwin, the naturalist, says of the inhabitants of Tierra del Fuego, that while his own party, who were well clothed and sitting close to a fire, were far from too warm, these naked savages, who were farther off, were observed, to his great surprise, 'to be streaming with perspiration at undergoing such a roasting.'" More likely, though, he had Emerson's essay "Self-Reliance" in mind when he wrote "Barbarism and Civilization." Emerson complained about excessive culture, arguing that "society never advances. . . . It undergoes continual changes; it is barbarous, it is civilized." Emerson believed that periodic returns to a more primitive life were invigorating, that infusions of brute vigor countered the draining effects of effete society. Contrast the "well-clad, reading, writing, thinking American," he said, with the "naked New-Zealander." Then compare "the health of the two men, and you shall see that his aboriginal strength the white man has lost."

Higginson disagreed. He spoke now on behalf of refined culture, of intellectual sophistication, of developed arts and letters. He told his Concord audience that cultured societies were in fact *more* robust and vigorous, and he attacked the "latent distrust of civilization" so prevalent in American society, which was still rooted in the frontier experience. There were misguided people who supposed that "refinement and culture are to leave man at last in a condition like that of the little cherubs on old tombstones, all head and wings." But this was an error. "Savage tribes," Higginson said, were in fact physically smaller, weaker, and more likely to succumb to illness. They were "always tending to decay."

Higginson considered it unnecessary to define labels we now consider invidious: *civilization, barbarism, savagery.* At the time, even the most enlightened New England abolitionists believed that black- and brown-skinned people inhabited a lower rung on the ladder of civilization. What made Higginson's use of these categories noteworthy, however, was his insistence that race had little to do with them. Civilization and skin color were unrelated, he claimed, "for the most degraded races seem never to be the blackest, and the builders of the Pyramids were far darker than the dwellers in the Aleutian Islands." In fact, he continued, the "black man apparently takes more readily to civilization than any other race."

As it turned out, terms such as *barbarism* and *civilization* had long been deployed in the debate over slavery. The *New York Times,* commenting on Missouri's decision to expel free blacks from the state, lamented, "Nothing has ever told so sad a tale of the relapse of the Slave States towards barbarism." The poet William Cullen Bryant had recently observed that Kansas was "a potent auxiliary in the battle we are fighting, for Freedom against Slavery; in behalf of civilization against barbarism." Higginson's lecture was a thinly veiled attack on slavery, which he considered truly savage. If Americans had engaged in the most astonishing act of civilization in history, he now told his Concord audience, "there yet lingers upon this continent a forest of moral evil more formidable, a barrier denser and darker, a Dismal Swamp of inhumanity, a barbarism upon the soil, before which civilization has thus far been compelled to pause."

The lecture was about John Brown after all.

Henry David Thoreau had also attended Higginson's talk. As soon as it was over, he walked with Alcott along the Lexington Road to the white clapboard house that belonged to their mutual friend, Ralph Waldo Emerson. Emerson was gone—his annual lecture tour had taken him to western New York—but the rest of his family was present, as well as several members of the Concord Ladies' Anti-Slavery Society. Earlier that day these women, including Alcott's wife and his daughter Louisa, had hosted a tea for Higginson in appreciation of his support of John Brown. Now the small group gathered in the parlor and listened while Alcott and Thoreau argued over Higginson's lecture.

Thoreau was nettled. He had studied Native Americans for decades, had read accounts of their customs and lore, had collected their artifacts. He had tried, in his way, to live as they did. He was convinced that native peoples filled a crucial niche in creation, bridging the gap between nature and civilized society. Like Emerson, he believed that society threatened the vigor of the individual. The tonic to this condition was wilderness: a return to those unmediated conditions from which humans had originally sprung. Thoreau not only celebrated raw nature, but also admired those peoples who still lived in close contact with it.

Alcott understood the lecture as an implicit repudiation of "Thoreau's prejudice for Adamhood." Too often, he thought, his friend celebrated physical reality at the expense of the spiritual. Alcott agreed with Higginson: civilization was "the ascendency of sentiment over brute force, the sway of ideas over animalism, of mind over matter." If human history was progressive, an upward movement toward gradual perfection, then Thoreau was wrong to sympathize with native peoples. After all, they represented a step backward. Alcott challenged Thoreau's identification "with the woods and the beasts, who retreat before and are superseded by man and the planting of orchards and gardens. The savage succumbs to the superiority of the white man."

Alcott was not particularly interested in Higginson's point about race. He was too busy fighting a battle against reductive materialism. Thoreau's praise for aboriginal peoples annoyed him primarily because it valued the animalistic aspects of human nature instead of its loftier aspirations. This way of thinking was exemplified in the new book by Darwin, which kept coming up in Concord conversations that winter. *On the Origin of Species* had penetrated the circle of Concord transcendentalists as quickly as it had swept through Boston, with Thoreau mentioning it to Emerson sometime earlier that month. Emerson promptly included it in a list of books he wished to read, and in February, while still on the road delivering lectures to the raw western states, he wrote from Lafayette, Indiana, to his wife, Lidian, to ask that she acquire a copy. Emerson felt like "an old gentleman plodding through the prairie mud," he wrote, adding, "I have not yet been able to obtain Darwin's book which I had depended on for a road book. You must read it,—'Darwin on Species.' It has not arrived in these dark lands."

Emerson was always interested in new books and new ideas, but this one was of special interest because it corresponded to a pattern of thought to which he was already predisposed. For decades he had relied on a quasi-evolutionary language to convey his sense that existence was an unfolding process toward perfection. In the 1830s, when his most daring and original ideas had first occurred to him, he eagerly consumed any book that confirmed those ideas, including the scientific theories of Lyell and Lamarck, whose notion that the Earth's crust perpetually remade itself he found particularly congenial. A decade later Emerson appended a poem to *Nature,* inspired by a diagram he had seen in Alcott's journal. Describing "a subtle chain of countless rings," Emerson imagined a procession of development in which

> striving to be man, the worm
> Mounts through all the spires of form.

In 1860 Emerson was revising a manuscript that married this evolutionary thinking to a bleaker materialism. *The Conduct of Life,* published at the end of the year, is a somber book, as pensive and stern as a Doric pediment from a Greek temple. No longer a young visionary, Emerson now directed his attention to those biological and social limitations experienced by every person. "Nature is no sentimentalist," he wrote. "The habit of snake and spider, the snap of the tiger and other leapers and bloody jumpers, the crackle of the bones of his prey in the coil of the anaconda,—these are in the system, and our habits are like theirs." Emerson would remain interested in the ways human intelligence might outwit the determining parameters of biology, but he held few illusions about our ability to escape limitations entirely: "The face of the planet cools and dries, the races meliorate, and man is born. But when a race has lived its term, it comes no more again."

Bronson Alcott was all too familiar with such misguided ideas. Fifteen years earlier, when the anonymous *Vestiges of the Natural History of Creation* arrived in America and offered its grand hypotheses about nebular fires and the evolution of man from monkeys, he had scrupulously read it, hoping to gather insights into the inner workings of the Creator. He even discussed the book in his "Conversations," praising its use of science to illuminate the divine plan. He agreed with the author that creation rests "on one law and that is,—DEVELOPMENT"; after all, like most of his transcendentalist friends, Alcott was convinced that life was destined toward perfection.

But that was as far as he could go with the materialist philosophy contained in *Vestiges.* As he had once told young Franklin Sanborn, the problem with science in general was that it threatened to reduce the world to pure phenomena—to *stuff.* "Naturalists . . . begin with matter," he complained, when "they should begin with spirit,—as in the 'Vestiges' the author supposes man developed as a final product from inorganic matter. This is wrong," Alcott insisted. ". . . Matter is the *refuse* of spirit, the residuum not taken up and made pure spirit."

Alcott was quarreling with science's commitment to materialism, its dedication to describing physical laws that could be objectively verified. But he was also making a claim about the spiritual nature of the universe. For Alcott, the world rendered visible by scientific analysis was ultimately a closed system, a walled fortress that rebuffed any fact or law that could not be empirically measured. Yet God was the most important fact of all. Most antebellum scientists thought they were engaged in the study of His creation, but for Alcott they had already disqualified themselves for such study by restricting their work to purely material phenomena.

It was as though Darwin were looking through a telescope from the wrong end, focusing on the solid and visible aspects of creation instead of examining the truly marvelous spirit that infused all matter and endowed it with meaning. Spirit was the key, he once told Sanborn; spirit permeated everything. "It is like a swarm of bees," he tried to explain. "They are *conical,* like the arrangement of *things* and man. All the bees depend on the queen bee; so all *matter* depends on man."

This image understandably confused Sanborn, so Alcott tried another approach: "It is better to say boldly that we are not formed from matter, but that we ourselves form it." As the eye in essence *creates* what it looks upon, so too is matter created by human consciousness. Sanborn admitted that this might be "nearer the truth." Perception, after all, shaped each individual's world to some extent—but did not the world also shape our perception? Not according to Alcott, who believed that the divine spirit poured through each individual, enabling him or her to fashion the material universe in a creative act. "Mr. Alcott," Sanborn noted with some exasperation, "seemed to imply [that this] was almost the *exact* truth."

Alcott was making a similar point now in Emerson's parlor. "The more animated the brain," he announced, "the higher is the man or creature in the scale of intelligence." On the other hand, the "barbarian has no society." Thoreau's quarrel with civilization made sense only if society was incapable of further development. Of course, there was "no civilized man as yet, nor refined nations, for all are brute largely still." But Alcott was confident that things were changing, that in the future people would shed their animalistic urges, transcend the degradations of the body, and become pure spirit. "Man's victory over nature and himself," he told the group assembled in Emerson's parlor, "is to overcome the brute beast in him."

At some point Thoreau excused himself from the conversation. In *Walden* he admiringly called Alcott one "of the last of the philosophers,— Connecticut gave him to the world,—he peddled first her wares, afterwards, as he declares, his brains." Thoreau respected Alcott's habitual faith that "a better state of things than other men are acquainted with" did in fact exist. But now he remembered the library book he had checked out earlier that day. It sat on the desk in his third-floor attic room, its spine unopened, its pages uncut. It dealt with facts and observations, not abstractions, and it beckoned like a siren. After a short walk to his mother's house on Main Street, he climbed the stairs to the attic, pulled out the large natural history notebook from his bookshelves, and began transcribing passages from the new work.

7

The Nick of Time

Henry David Thoreau was forty-two years old when he encountered Charles Darwin's *On the Origin of Species*. Short and wiry, agile and athletic, his two most prominent features were an enormous nose and penetrating blue eyes. In the few photographs we have of him, these eyes alternate between a dreamy inward softness and a cold, crystalline intensity. The contrary expressions hint at the transcendentalism he discovered in college and the rigors of science that began to compete for his imagination a dozen or so years later. It was typical of him—and of his era—that the two impulses shared space inside the same head.

His journals from the period reveal him on a typical winter day in 1860: sliding, lurching, tramping through the snow in a pitch pine forest outside Concord, studying the tracks of a partridge, sampling a shriveled chokecherry, then stepping onto the ice of the Assabet River to watch the coppery minnows dart in all directions below his feet. Every day of every season, Thoreau descended from the third-story attic of his mother's renovated yellow house on Main Street, put on a battered hat, and spent at least four hours sauntering (his word) through the woodlots and forests and pastures that surrounded Concord. Like a window-shopper strolling the boulevards of Paris, he examined flora and fauna, bending to inspect a robin's egg in the spring, pausing to taste a huckleberry in the late summer. He constructed a platform inside the crown of his hat so that he might store the botanical samples he collected on these walks. (One autumn, when he was carrying home the seedpods of touch-me-nots, their explosions sounded like pistol fire above his head.) In the evenings he returned with these treasures to his attic study, where he transcribed his observations into a succession of notebooks stored in a box he had built especially for the purpose.

Born into a modest family that operated a local pencil-making business, Thoreau became recognizably himself—the cantankerous part-time hermit and philosopher-naturalist—after reading the works of his fellow villager Ralph Waldo Emerson, while still a student at Harvard. The precise circumstance of their first meeting is unknown. One account has Thoreau's sister sending the eminent writer copies of her brother's poetry. Another has Thoreau walking to Boston and back, a journey of some forty miles, to hear Emerson lecture. What we know for certain is that sometime in the fall of 1837 the two became inseparable. Emerson noted in his journal, "My good Henry Thoreau made this else solitary afternoon sunny with his simplicity and clear perception." Thoreau began to keep a journal shortly after meeting Emerson. His first entry: "'What are you doing now?' [Emerson] asked, 'Do you keep a journal?'—So I make my first entry to-day."

Much has been made of the complicated, intense, and sometimes rivalrous relationship that soon developed between the two. Both Emerson and Thoreau idealized friendship, yet both were prickly, awkward, and aloof. Emerson admired Thoreau's deep knowledge of nature and his simple practicality, so unlike his own cerebral refinement. In 1841 he invited Thoreau to live in his home. Later that decade Thoreau moved in again, this time to care for the Emerson family while the older man traveled through Europe on a speaking tour. For his part, Thoreau identified with Emerson to such a degree that he became the subject of local ridicule. One contemporary, writing in 1848, snidely remarked that he was "all overlaid by an imitation of Emerson; talks like him, puts out his arm like him, brushes his hair in the same way, and is even getting up a caricature nose like Emerson's."

There were tense moments. Emerson complained that "Henry does not feel himself except in opposition" and remarked, "As for taking Thoreau's arm, I should as soon take the arm of an elm tree." Thoreau ever felt his mentor's forceful personality. "Talked, or tried to talk with R.W.E.," he bitterly observed. "Lost my time—nay, almost my identity." In 1857 he briefly felt as though the association had run its course: "And now another friendship is ended. . . . With one with whom we have walked on high ground we cannot deal on any lower ground ever after. [Emerson and I] have tried for so many years to put each other to this immortal use, and have failed."

Still, the relationship frequently brought out the best in both writers. When Emerson announced in 1841 that "part of the education of every young man" should be to put himself "into primary relations with the soil and nature," he was thinking of Thoreau, whose passion for nature seemed an antidote to America's rampant materialism. Prompted by the statement, Thoreau moved to a parcel of land Emerson owned near Walden Pond. In the most celebrated act of his career, he built a rough-hewn shack seventy yards from the shore and commenced his famous experiment to "live deep and suck out all the marrow of life, to live so sturdily and Spartan-like as to put to rout all that was not life." Rising at dawn and bathing in the pond's glassy water each morning was "a religious exercise," an opportunity to become awakened each day by "newly-acquired force and aspirations [to] a higher life than we fell asleep from."

The two years at Walden made him a writer. It afforded him a writer's most precious commodity: time. Each day he sat at the little table in his cabin and worked steadily on a book that would eventually become *A Week on the Concord and Merrimack Rivers,* a pastoral eulogy to his older brother John. Thoreau idealized his brother, who was thought by many in Concord to have been the more talented and certainly the more extroverted Thoreau son. The brothers went hiking together; they fell in love with the same woman. When Henry graduated from Harvard, they founded a short-lived school together. In 1841 John began to show signs of tuberculosis, an often-fatal illness that affected as many as one in four of all New Englanders. But it was something else that killed him. Stropping a rusty razor, he nicked his ring finger and contracted lockjaw. He died, delirious, in front of Henry, who did not outwardly mourn but soon began to exhibit symptoms of lockjaw, too. Thoreau was seized by convulsive spasms; he grew feverish and his jaw stiffened. The family gathered around his bed and prepared to say goodbye to their only surviving son. Gradually, however, the symptoms disappeared. The illness had been psychosomatic, a sympathetic response to his brother's suffering. For years, Thoreau's eyes watered whenever his brother's name was mentioned.

Two years on Emerson's woodlot would provide him with the material for his masterpiece, *Walden,* a book still cherished by countless readers who admire its nonconformist vision and evident love of nature. All of us dream of escape from our quotidian lives, and Thoreau's decision to sequester himself in the woods can seem, especially in our rushed and hyperconnected world, an act of heroic defiance. Yet many readers forget that his escape to the woods was in service to bigger, more pressing questions. What was the best way to live? How should we spend the precious time allotted to us? In his hands, these problems became plot elements in a thrilling drama. "In any weather, at any hour of the day or night, I have been anxious to improve the nick of time," he declared, expressing the profound urge many of us feel to make our lives meaningful and worthwhile. Sitting outside his cabin while a chorus of bullfrogs or Canada geese filled the air, observing the subtle, rippling changes in the color of the pond—these pastimes were part of a larger project to fully inhabit the nick of time, "the meeting of two eternities," as he called it, "the past and future, which is precisely the present moment."

Transcendentalism would always appeal to Thoreau's defiant personality. The movement was not just an effort to invoke the divinity within. It was also a cultural attack on a nation that had become too materialistic, too conformist, too smug about its place in history. Since 1800 the United

States had grown rampantly, recklessly, improbably fast, its population doubling every two decades. By 1860 there were thirty-five million people in the nation, more even than in Great Britain. This unprecedented growth was in part a product of the country's abundant land and natural resources. And it was accelerated by the most transformative technological revolution in history to that point, a series of new inventions that made America the first truly modern nation. The daguerreotype, an early form of the photograph, allowed viewers to encounter people and places they had previously only read about; the telegraph similarly demolished old conceptions of space and time. But the most startling and marvelous new technology of the era was the railroad. In 1860 some thirty thousand miles of railroad track knit the nation together, linking isolated hamlets and encouraging mobility among a people who previously had seldom traveled beyond their county seats. For many, the train stood as an emblem of progress, democracy, and material wealth.

Technology and natural resources also combined to produce the single most important global commodity in nineteenth-century America: cotton. Thanks to the Massachusetts inventor Eli Whitney's cotton gin, the staple almost single-handedly propelled the United States onto the world economic stage. By 1850, American cotton accounted for 75 percent of the 800 million pounds consumed annually in Britain, and 90 percent of France's nearly 200 million pounds. Much of that cotton was spun in textile mills in Waltham and Lowell, two towns just outside Boston, which quickly became the throbbing heart of the Industrial Revolution in the United States. Thousands of young women—and many children—operated spinning machines and enormous looms powered by steam, often working fourteen hours or more a day. By 1860, these two mills alone operated more than five million spindles and spun a million bales of cotton annually.

What made this industry possible, of course, was slave labor. In 1860 some 85 percent of cotton picked in the South came from large plantations, where 90 percent of all the slaves in the United States lived and worked. Millions of slaves painstakingly harvested cotton bolls by hand, rising before dawn and filling their sacks until evening. To remove seeds, slaves operated large gins based on Whitney's patent; they used presses to shape the cotton into compact bales. The growing demand for cloth produced a corresponding demand for slaves, and by 1860 nearly four million slaves worked under coercion in what slaveholders proudly called the kingdom of cotton. Slavery was woven into the fabric of American life, the economies of the North and South, like nothing so much as the warp and weft of cotton threads in valuable gingham and calico. "You dare not make war upon cotton," exclaimed the governor of South Carolina in 1858, referring to rising abolitionist sentiment in the North; "no power on earth dares to make war upon it."

The transcendentalists came neither quickly nor easily to the abolitionist cause. In his most famous essay, "Self-Reliance," published in 1841, Emerson fulminated against social activists who wore their causes like the latest fashion: "If an angry bigot assumes this bountiful cause of Abolition, and comes to me with his last news from Barbadoes, why should I not say to him, 'Go love thy infant; love thy woodchopper: be good-natured and modest: have that grace.'" Emerson was being deliberately provocative, but he also believed that individuals had to undergo moral regeneration before turning their attention to social improvements. One could not reform the world before reforming oneself.

History soon challenged that assumption. Throughout the 1850s a series of disastrous political concessions revealed all too clearly that the liberation of the soul—especially the souls of enslaved African Americans—would arrive less quickly than Emerson had hoped. Repulsed by the "filthy enactment" of the Fugitive Slave Act of 1850, which he found hard to believe had been "made in the 19th Century, by people who could read and write," Emerson defiantly announced, "I will not obey it, by God." Next the infamous Kansas-Nebraska Act of 1854 did away with the Missouri Compromise's northern limit on slavery, allowing new territories north of 36 degrees 30 minutes to choose whether to become slave states. This led to the eruption of violence in Kansas between pro- and antislavery settlers, followed by John Brown's attack on Harpers Ferry and his subsequent hanging.

Emerson's journals throughout the 1850s are interspersed with expressions of outrage. As political tensions escalated, he urged dismantling the Union so as to protect the freedom valued by the North. Slavery was a communicable disease, a contagion. "We intend to set & to keep a *cordon sanitaire* all around the infected district," he wrote, "& by no means suffer the pestilence to spread." Publicly, he was just as adamant in his condemnation of slavery. "I do not see how a barbarous community and a civilized community can constitute one state," he announced to lyceum audiences. "I think we must get rid of slavery, or we must get rid of freedom." Political compromise was morally repugnant, contrary to logic: "We have attempted to hold together two states of civilization: a higher state, where labor and the tenure of land and the right of suffrage are democratical; and a lower state." The results had been disastrous.

If anything, Thoreau was even more extreme in his response to slavery. On July 23, 1846, while still living at Walden Pond, he walked into town and was accosted by Sam Staples, the local tax collector. Thoreau was six years delinquent on his poll taxes. He refused to pay because he opposed the Mexican-American War, believing it was being waged to expand slavery throughout the Southwest. As he later wrote, "I cannot for an instance recognize that political organization as *my* government which is the *slave's* government also." Staples had no other recourse but to put Thoreau in jail, where he spent a night before he was bailed out by someone he never identified but who was most likely his aunt.

Thoreau was powerfully affected by the John Brown affair. He kept pencil and paper under his pillow in order to write down his thoughts in the middle of the night. Three days after learning of the failed attack on Harpers Ferry, he came down to breakfast. His mother, sister, and aunt, all prominent local antislavery activists, were already seated. Thoreau asked them a question. He had been up all night, pacing in the attic, arranging his scattered notes into a lecture that would portray the captured man in a more just light. Now he wanted his family's advice. Should he speak publicly on behalf of Brown? To do so was dangerous. In the South, abolitionist sentiment was banned from newspapers and punished by law. Antislavery advocates were routinely killed or driven out of the region. In the North abolitionists were still sometimes attacked. They were tarred and feathered, publicly tortured. Sanborn had cautioned him against going public, conceding that the threat of physical assault was real. At the minimum, Thoreau would be pilloried in the press for his opinions.

The family voted two to one in favor of his speaking out.

A week later he gave a public address at the First Church in Concord. To those who warned him against defending a traitor, he replied, "I did not send to you for advice, but to announce that I am to speak." Thoreau's address was entitled "A Plea for Captain John Brown," and it became an instant sensation. Edward Emerson, the fifteen-year-old son of the transcendentalist, remembered how "deeply stirred" was Thoreau's voice on that occasion. He read his paper, Edward recalled, as if it "burned him." Bronson Alcott called the address "a revolutionary Lecture . . . by which the martyr's fame will be transmitted to posterity." Even Sanborn, who considered Thoreau a mediocre speaker, admitted that he had been "mightily stirred by the emotions" the address raised.

Thoreau repeated the address in Boston to more than two thousand people. (He filled in for Frederick Douglass, who had fled the country in the wake of the Brown affair.) Next he gave the speech in nearby Worcester, prompting local newspapers to pun on his name, calling him "a thorough fanatic," and to wryly note that Brown's trial "seemed to have awakened 'the hermit of Concord' from his usual state of philosophic indifference." Cumulatively, these speeches were decisive in turning the tide of opinion toward Brown among Northern abolitionists. They gave others the courage to speak on behalf of the executed man. Emerson soon echoed Thoreau's defense by comparing the gallows erected for Brown's execution to the cross.

On December 2, 1859—the day Brown was executed—Thoreau helped arrange a memorial service with Emerson, Alcott, and Sanborn. The day was unusually warm, as if the baleful news from Virginia had been carried on a spring breeze. Two hundred people gathered in the Town Hall to listen to the solemn eulogies. Civic leaders refused to permit the bells of the First Church to mark the occasion, but Alcott believed "it was more fitting to signify our sorrow in the subdued way, and silently." The choir sang a hymn; Alcott and Emerson read selections and poems by Schiller, Wordsworth, Tennyson, and Tacitus. Then Thoreau stood up. "So universal and widely related is any transcendent moral greatness—," he said to the assembled audience, "so nearly identical with greatness every where and in every age, as a pyramid contracts the nearer you approach its apex—that, when I now look over my commonplace book of poetry, I find that the best of it is oftenest applicable, in part or wholly, to the case of Captain Brown."

This is Thoreau the passionate abolitionist—almost as iconic as Thoreau the nature lover and advocate for simplified living. Not nearly so well known is Thoreau the scientist. Sometime in the early 1850s, several years before *Walden* was published, Thoreau began a new career and a new identity. He had always been obsessed with the workings of nature, with identifying new birds or plants. He delighted in the smell of wet stones, the rich tang of freshly spaded loam, the droning sound of the bumblebee as it stumbled drunkenly from flower to flower. Under the tutelage of Emerson and the emerging philosophy of transcendentalism, he had come to understand these delights as embodied revelation, living scripture. Nature was the expression of a benign divinity that communicated spiritual truths to the solitary individual.

While these insights helped explain the thrill and pleasure the young Thoreau felt in the woods of Middlesex County, they didn't entirely answer the demands of his personality, which was congenitally skeptical and searching. He was drawn to the concrete and the palpable, to the heft and texture of experience. Increasingly he relied on the tools of science to make sense of the woodlots and ponds to which he daily tramped. He measured and weighed, touched and tasted. He carefully wrote down his observations.

Occasionally he worried about this new tendency. Focus on the particular instead of the abstract seemed to him a sign of age. A "young man is a demigod," he confessed in his journal, comparing his youthful raptures with the prosaic realities of "the grown man, alas! [who] is commonly a mere mortal." "I fear," he wrote on August 19, 1851, that "the character of my knowledge is from year to year becoming more distinct and scientific; that, in exchange for views as wide as heaven's scope, I am being narrowed down to the field of the microscope." On Christmas Day 1851 he put the problem to himself more tartly: "What sort of science is that which enriches the understanding, but robs the imagination?"

One night he was awakened by a dream. Astride two fractious, ungovernable horses—literal nightmares—he galloped through the woods. "In my dream I had been riding," he wrote, "but the horses bit each other and occasioned endless trouble and anxiety, and it was my employment to hold their heads apart." He didn't bother to analyze the dream, but the tugging horses suggest someone trying to coax the contending forces of his life into a shared direction. Daybreak brought no resolution to the problem: "I awoke this morning with infinite regret."

The dream had come several months after Thoreau read Darwin's stirring travel memoir, *The Voyage of the Beagle*, first published in America nearly a decade earlier. He had loved the book from the first. His natural history notebook was soon filled with extracts from the *Voyage* on the tiny aquatic creatures known as infusoria, on palm trees and coral atolls, on the way Argentinians hunted partridges, on the size of hail in Buenos Aires, on the invasive species of fennel introduced to the New World from Europe, on the tooth of a prehistoric horse found by Darwin on the Pampas, and on the habits and features of llamas. Less than a week after finishing the book, he opened it up and read the whole thing over again, liberally copying from it once more. He appreciated Darwin's disparate interests, his omnivorous attitude toward knowledge, his reliance on close observation. He considered the English explorer a colleague of sorts: "His theory of the formation of the coral isles by the subsidence of the land appears probable."

Darwin's travelogue seems to have encouraged Thoreau to begin his own idiosyncratic voyage of discovery. Instead of circumnavigating the globe, he ventured each day into the woods and fields encircling Concord—weighing, measuring, and categorizing the local flora and fauna. He carried with him a small diary and a pencil, a spyglass for observing birds, a portable microscope, and a jackknife. He also carried two books: Asa Gray's *Manual of the Botany of the Northern United States* (invaluable for its line drawings and comprehensive verbal depictions of plant genera and families) and a large and cumbersome book of flute music entitled *Primo Flauto* that had belonged to his deceased father. He used this volume to press plants and flowers for his burgeoning herbarium.

Like Darwin, he took notes: thousands and thousands of pages of them. In less than a decade he produced a series of notebooks filled with drawings and descriptions of mushrooms, animal tracks, snowballs, leaves, and flowers. He meticulously described the blooming seasons of plants, the subtly changing colors of leaves, the ripening of berries. He measured the heights of grasses and the size of red maple leaves in May. What he intended to do with all this data is still not entirely clear. Scholars believe he may have envisioned a vast, encyclopedic calendar of Concord's seasons, a hybrid text of empirical scrutiny and poetic observation unlike anything else in American literature. It seems probable that he was trying to yoke the fractious horses of his dream, to straddle transcendental idealism and scientific empiricism, but he was evasive about his plans. Describing his research to a young Michigan schoolteacher in 1856, Thoreau sheepishly confessed, "I am drawing a rather long bow."

What stands out most from this period is the pagan joy coursing through every word in his journal: a relish and delight wholly unrelated to transcendental raptures. Thoreau admires the gossamer filaments that glisten in the sun when he tears apart a milkweed pod. He samples the bitter juice of unripe berries or amuses himself by measuring his strides as he slides across frozen rivers. In 1857 he began an elaborate series of experiments to discover the best way to boil tree sap into syrup. Later he attempted to make wine from birch bark. His interests branched apart, proliferated, carved new channels of thought. He delved into cartography and the magnetic variations of compasses. He studied geology. By 1860, his third-story attic room had become a private natural history museum, stuffed with birds' nests, arrowheads, and more than a thousand pressed plants. On shelves made from driftwood he had gathered at Cape Cod, he kept the skins of reptiles, assorted pelts, rocks and stones, lichens, moss, and the carcass of a Cooper's hawk as well as its spotted bluish-white egg.

Early in the twentieth century, the nature writer John Burroughs, an unabashed Thoreauvian, thought he discerned a problem in all this. Thoreau's journals were simply too capacious, too indiscriminate, guided by no principle of selection. They were "a hungry, omnivorous monster that constantly called for more." Bedazzled by the lush plenitude of nature, Thoreau had difficulty organizing his material into a coherent project. As he put it himself in *A Week on the Concord and Merrimack Rivers,* "Observation is so wide awake, and facts are being so rapidly added to the sum of human experience, that it appears as if the theorizer would always be in arrears."

Put another way, he had adopted the methods of science without the benefit of a scientific theory.

Books can change a life. The nineteenth century is filled with accounts of readers who opened a novel or a collection of poetry and instantly became someone else. Moncure Daniel Conway, a young Virginian idling on his father's plantation, opened a copy of Emerson's *Essays* one morning and stumbled across a single sentence that sparked a complete transformation of his being. He left the South and repudiated slavery. (He also refused to divulge which sentence effected the transformation.) The great orator and author Frederick Douglass experienced something even more profound. At age twelve he came into possession of a popular collection of political dialogues entitled *The Columbian Orator,* "choice documents to me," he recalled. "I read them over and over again with unabated interest. They gave tongue to interesting thoughts of my own soul, which had frequently flashed through my mind." From that book Douglass came to realize the injustice of slavery—in particular the injustice of his own enslavement. The book served as a base for his future life. It prompted him to escape to freedom, and it influenced his career as an abolitionist speaker for decades to come.

Books can also change the world. In 1852 *Uncle Tom's Cabin* crashed through American society, cracking its foundations and altering the era. Harriet Beecher Stowe's first novel was an immediate sensation. Readers throughout the nation, both North and South, found themselves captivated by the tale of Tom and little Eva, often with an intensity they had never felt before. Some readers experienced guilt at having succumbed so completely to the work's imaginative power. "I have indulged myself," wrote one reader with puritanical shame. "I was bound down captive," admitted another. Mary Pierce Poor, of Massachusetts, so feared the novel's ability to enchant a sick relative that she wrote home "to advise" her "not to attempt [*Uncle Tom's Cabin*]. I am afraid it would kill [you]. I never read anything so affecting in my life."

Something like this happened to Thoreau when he checked out the *Origin* from the Concord Town Library. This copy was not the same edition Darwin had sent to Asa Gray but rather an unauthorized reprint from the New York publisher D. Appleton, bound in a grayish-brown cloth with gilt lettering on the spine. Thoreau read this edition in late January 1860, bright warm lamplight spilling onto his desk and across the book. He must have experienced that loss of self so many readers undergo when they launch into the secret world of their favorite volume. As he turned the pages, a new universe took form on the rectangular page before him. He took notes in the large natural history journal he kept opened on his desk.

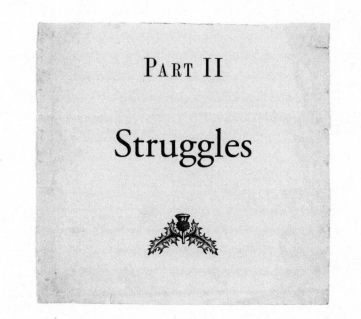

PART II

Struggles

8

Bones of Contention

Early in 1860 members of the Boston Society of Natural History began to hear rumors of a remarkable collection of gorilla specimens that had arrived from the Gold Coast of western Africa. These rumors were spread in part by Jeffries Wyman, Harvard's comparative anatomist, who was president of the society and eager to share information about the creature. In the close-knit, cloistered world of Boston science, word quickly spread, and on the fourth day of the year, in the stolid stately building on Mason Street, the society's lecture room hummed in anticipation.

It is one of those remarkable historical coincidences that Darwin's book on evolution reached America weeks after John Brown's execution and just as the first gorilla specimens were crossing the Atlantic—ensuring that the issues of slavery, primate descent, and evolution would be inextricably linked in the minds of the *Origin*'s first readers. Later in the century, and especially after the Civil War, Darwin's book would exert enormous influence on American thought; it provided an ordering principle for a society that seemed to grow more complex each year. Sociologists soon applied Darwinian theory to society, suggesting that people who were best adapted to their environment would invariably flourish. (Social Darwinism was primarily a rationale for preserving the status quo.) But in 1860, whenever Darwin's book was discussed and debated in lecture halls and newspaper columns, it was almost always discussed and debated as an important commentary on racial issues that had long smoldered at the heart of American society.

Members of the Boston Society of Natural History were also interested in the book's *science*. The organization had been founded thirty years earlier, conceived as a gathering place for civic-minded physicians and amateur naturalists who wanted to display their collections of seashells and rocks and stuffed birds and to discuss the latest developments in geology, natural science, and medicine. By 1860, it contained the nation's second-largest collection of objects from the natural world (the largest was in Philadelphia), with some 257 mammalian crania, four thousand shells, and countless herbaria. It boasted a fine library of five thousand volumes and a membership that included Asa Gray, Ralph Waldo Emerson, and Henry David Thoreau, who was invited to be a corresponding member in 1850 and who thereafter contributed occasional plant and animal specimens. (In 1860, he donated a stuffed lynx.) Other prominent members included Dr. Benjamin Joy Jeffries, an ophthalmologist, whose grandfather had been the first person to cross the English Channel by aerial balloon. (He averted a fatal crash by casting off all ballast, including his clothing, into the sea below the towering Pas de Calais.) Another prominent figure was the society's vice president, Dr. Charles T. Jackson, who was Emerson's brother-in-law. Brilliant and mercurial, Jackson was a physician, geologist, and inventor who had studied at the Sorbonne and was the proud owner of the most advanced private laboratory in Boston.

Darwin's book was not on the agenda at the January 4 meeting of the Boston Society of Natural History, but members discussed it that evening anyway, presumably after Wyman described his recent trip to New York. There he had visited the "Gorilla collection of Mr. Du Chaillu . . . made during a residence of three or four years in the country of this largest anthropoid ape." Paul du Chaillu was the French-born explorer who had supplied Wyman with several gorilla pelts and skeletons a few weeks earlier. In October he had left Senegambia and traveled to Manhattan, where he hoped to make his fortune displaying African artifacts. Du Chaillu stayed at the Five Points Mission (he was practically destitute) and opened a modest exhibition hall on Broadway, where he displayed spears and shields from various tribes within the African interior and brightly colored birds he had shot in equatorial jungles. "But the guardian genius of the place," according to a reporter for the *New York Post,* "is the gorilla, or man-monkey—one of the troglodyte tribe." In the center of the exhibition space were mounted a series of stuffed gorillas in various threatening poses. Du Chaillu stood before these creatures and lectured, debunking rumors of their intelligence and ability to fight rivals with clubs and other weapons. Nevertheless, he insisted, the gorilla was "among the most formidable animals, its arms being as large as some men's legs."

Wyman announced to members of the society that du Chaillu had generously donated several gorilla specimens to Harvard in the name of science. But his talk focused on the exhibit in New York, which contained "an extensive series from the quite young to the adult animal." Of the fourteen specimens on display, ten were male. Wyman had closely examined them, and he had inspected the skull of a female, discovering that "the teeth were in a continuous series as in man, whereas in all the anthropoid monkeys there is normally an interval between the upper incisors and the canines."

While in du Chaillu's exhibition hall, Wyman did what every anatomist of his generation would have done: he measured the volume of each creature's skull. The process was simple; one loaded an empty skull with shot, sand, or some other substance to determine how many cubic inches of space were filled. Craniometry, as it was called, had begun in the late eighteenth century as a way to determine basic differences between human and animal skulls. By the mid-nineteenth century the science was increasingly employed to determine discrepancies in the cranial capacity of different races. The assumption was that brain size indicated intellectual capacity, which in turn led to cultural development. Measuring the skulls of whites, blacks, Native Americans, and people of other races quickly became a cottage industry of American science.

The gorilla's cranial capacity ranged "in the males from 24 to 34 ½ cubic inches, the average being 28 or 29," Wyman reported. Because the gorillas weighed more than humans, "the relative size of the brain is very small; in the lowest human races the cranial capacity is about 75 cubic inches, and the weight of the body about 150 lbs." Still, there were structural similarities between humans and gorillas that suggested a startling relationship between the two. For example, "In the young gorilla, as in other anthropoid apes, there is no indentation of the body at the waist; the back also forms a regular curve, as in the human foetus or new-born child." Likewise, the ear of the gorilla "comes nearest to the human ear" of any other primate. Wyman thought the gorilla's "muscles are considerably modified from the human," yet most "of the muscles of the human hand are found in the gorilla," the one exception being that the primate's "thumb is supplied by a slip from the common flexor."

These descriptions were intriguing. On the one hand, Wyman's cranial measurements pointed to clear distinctions between humans and primates. On the other, anatomical differences between the two species were not as clear-cut as one might expect. Humans and gorillas shared many of the same physical structures, some identical, others slightly modified. If one conceded these physiological similarities, the next step was to consider whether they pointed to a relationship of some kind, a linkage uniting the two species. Although this went against biblical teaching, du Chaillu's gorillas offered compelling proof for such a linkage, even suggesting that a beast resided within each of us, its history lodged in our bones.

Du Chaillu would soon capitalize on the similarities between gorillas and humans when he published a book about his experiences entitled *Explorations and Adventures in Equatorial Africa.* The travel narrative met with immediate and widespread praise. "M. du Chaillu has struck into the very spine of Africa," observed the London *Times,* "and has lifted the veil of the torrid zone from its western rivers, swamps, and forests." But most readers were interested primarily in the author's account of gorillas. The creatures ran "on their hind legs," du Chaillu wrote, looking for all the world and rather "fearfully like hairy men; their heads down, their bodies inclined forwards, their whole appearances like men running for their lives." There was something indeterminate about du Chaillu's gorillas; they were simultaneously human and not human. "Take with this their awful cry, which, fierce and animal as it is, has yet something human in its discordance, and you will cease to wonder that the natives have the wildest superstitions about these 'wild men of the woods.'"

The uncanny human resemblance of gorillas most likely fueled the discussion of Darwin's *Origin* at the Boston Society of Natural History. We have no full account of that discussion, but we can piece some of it together from the letter Asa Gray wrote his English friend Joseph Dalton Hooker the following day. One of Gray's colleagues at Harvard apparently "had read but part of" Darwin's book. "He says it is *poor—very poor*!! (entre nous)." Rather than displeasing Gray, this news in fact made him giddy. "The fact he growls over it, like a well cudgeled dog,—is very much annoyed by it—" struck Gray with "great delight—."

Hooker soon forwarded Gray's letter to Darwin, who was also delighted. He asked, in the name of "pure vanity," to keep "the splendid, magnificent letter," and he relished the outrage of Gray's Harvard colleague. "It is rich," he wrote, "about Agassiz."

9

Agassiz

The man who would become the most emphatic critic of Darwinian theory in America had been born in Switzerland, the son of a strict provincial pastor. By age twenty-five Jean Louis Rodolphe Agassiz had clinched his reputation in Europe as a foremost ichthyologist, learning more about fish than anyone else in the world. Next he developed the Ice Age theory, arguing that the Earth had undergone repeated periods of glaciation: long frozen eras in which much of the Earth was covered in ice. During these periods, he said, entire families of plants and animals had been wiped from the face of the Earth while sublime marvels such as the Swiss Alps had been created.

By 1846, when he visited America on a lecture tour, Agassiz was no longer a productive scientist so much as an iconic representation of what science *looked* like. Witty and spellbinding, he lectured to workingmen's associations, lyceums, and women's salons about the wonders of nature. To a nation hungry for self-improvement, he brought recent scientific discoveries thrillingly to life. On the basis of these qualities, Harvard hired him, and by 1860 he was in the process of consolidating his formidable scientific reputation with an ambitious museum of comparative zoology. From the Commonwealth of Massachusetts he had extracted an unprecedented gift of $100,000 (well over $2 million in today's money), and the first floor of the new building on Oxford Street had just been completed. As he enjoyed explaining to potential donors, visitors would soon enter the museum through its impressive pillared portico and step into a spacious hall that revealed nothing less than the Great Plan of Nature herself. (Agassiz tended to speak in capital letters.) Eschewing traditional natural history cabinets, the museum would strive for unprecedented transparency in the arrangement of its collections. One wall would display the radiata—symmetrically identical animals such as starfish and sea urchins—while another would reveal the lower vertebrates, including birds and reptiles. The lower creatures would lead ineluctably to the Mammalia, the highest type of animal, Agassiz explained, "to which we ourselves belong." Although not yet completed, the exhibition hall would eventually culminate in a display "of men, skulls, skeletons, &c., for the study of the human races."

Agassiz's museum was predicated on the idea that science was nothing less than the study of God's creation. The universe was harmonious, unified, and governed by elegant laws. It was radiant with moral and aesthetic beauty. The Creator had stocked Earth with a delightful assortment of plant and animal species, each fashioned for its particular niche in the environment, each distinct and unlike any other. If the naturalist's job was to collect and describe these species, to catalog their wondrous diversity, such a task was in the service of a much nobler object. Agassiz studied nature because it revealed "the free conception of the Almighty Intellect, matured in his thought before it was manifested in tangible external forms."

He did not originate the concept of special creation—the idea that God had made each plant and animal species separately and distributed them across the planet according to divine logic. Nor was he the first to proclaim that species were therefore incapable of change. But Agassiz was far and away the most influential scientist to advocate for these ideas, arguing that every plant and animal had been meticulously fashioned by a divine providence and placed in a special "zone" for which it was perfectly suited. This was why koala bears are not found in the woods of

Vermont or whales in Lake Michigan. This was why tropical plants are so admirably matched to the humid environment in which they flourish.

Agassiz's special creationism was encouraged by his mentor, the great classificatory naturalist Georges Cuvier, but it also owed a debt to his father's religious teachings and, especially, to the German idealism that had inspired his youth. Early nineteenth-century thinkers such as Fichte and Schiller had asserted that God was the first and final reality. For them, the universe was a model of His mind. God had merely to conceive of stars or mammals or algae, and these things sprang into existence. Humans were unlike all other living things because they alone had been endowed with the capacity to appreciate the divine intelligence animating nature. Gray once tried to describe Agassiz's thinking in the following way: "There is order in the universe; that order presupposes mind; design, will; and mind or will, personality."

All this was enormously popular in Agassiz's adopted country, where it neatly corresponded to the reigning American philosophy of the day, transcendentalism. Agassiz preferred to discuss scientific matters with Emerson rather than with anyone in his own profession, and he agreed wholeheartedly with Emerson's idea that "behind nature, throughout nature, spirit is present." As he never tired of repeating, in a phrase that might have appeared in some transcendentalist manifesto, the "study of nature is intercourse with the Highest Mind."

In 1846, when Agassiz first arrived in America to deliver the Lowell Lectures in Boston, Asa Gray had served as his cicerone to the American scientific community, escorting him to Princeton, Philadelphia, and Washington. "He is a fine, pleasant fellow," Gray gushed to a friend, describing Agassiz's course of lectures as rich in learning and "planned on a very high ground." Back in Cambridge, where he was soon hired to direct Harvard's new Lawrence Scientific School, Agassiz paid regular visits to Gray's botanical gardens, perusing his colleague's herbarium while he puffed on his ubiquitous cigar. He is "a good genial fellow," Gray repeated a year or so later—but this time he hinted at troubles to come.

The two had evidently disagreed over some matter of botanical evidence. Agassiz, Gray wrote, "bears contradicting as well as—could be expected."

The tension between the two men was personal and professional. Agassiz was vivid and electrifying in the lecture hall, an extrovert who liked nothing so well as to speak to large, appreciative audiences. Gray was jealous of Agassiz's charisma, but he also distrusted the man's easy popularization of science. He complained that his colleague "is always writing and talking *ad populum*—fond of addressing himself to an incompetent tribunal." In his opinion, Agassiz exaggerated, simplified, and otherwise cut corners in order to entertain a nonprofessional audience.

Gray thought Agassiz's need for acclaim compromised his science. His early successes had been grounded in close observation—in the minute examination of fins and scales and prehistoric teeth, or of the jagged scoriae slicing through the alpine granite of his native country. Back then, hypothesis and observation had worked in delicate, contrapuntal relationship: ideas tested by experience and experience made comprehensible by ideas. But this supple habit of thought had apparently disappeared the moment Agassiz stepped onto the pier at Boston Harbor and began to cultivate his American reputation. He continued to catalog an ever-growing inventory of animal species—he even published a book or article now and then—but he no longer developed working hypotheses. His thinking had entered an ice age of its own: old ideas frozen in a kind of mental permafrost that hadn't thawed in decades.

If Gray was jealous of Agassiz's museum of comparative science, he also believed the institution had been conceived upon obsolete premises. Agassiz assumed his museum presented the architecture of creation. It was a museum of God's thoughts. And because God did not make mistakes, each indigenous species displayed in the museum was understood to be immutable. Gray believed this was factually incorrect. He had catalogued too many plant varieties to think each was separate and fixed. Sedges, mosses, and wildflowers exhibited countless varieties, one grading into the next. Present several botanists with the same evidence, and you would find them incapable of agreeing where one species stopped and another began.

Gray was especially disturbed by Agassiz's tendency to apply the logic of special creation to humans, which is another way of saying he was disturbed by the man's politics. Agassiz claimed that each race had been created independently in the zone best suited for it. White people were formed in eastern Europe, Native Americans in the western hemisphere, blacks in Africa. Surely this separate creation implied a hierarchy. God had placed blacks and whites, brown-skinned and yellow-skinned peoples in various locations for a reason, Agassiz argued, and in the process He had ranked them by their development. This argument made Agassiz not only the premier American scientist of his time—it also made him the nation's premier polygenist.

Ethnology—or the study of the origins, languages, and customs of various peoples—had arisen as a scholarly discipline in the late eighteenth century in response to the baffling diversity of peoples encountered by European explorers and merchants during the era of global expansion. American practitioners began to advance the discipline in the 1830s. A correspondent for the *Methodist Review* neatly summed up the science's core issue in 1844: "In surveying the globe in reference to the different appearances of mankind, the most extraordinary diversities are apparent to the most superficial observer." Two questions therefore presented themselves: "*Have all these diverse races descended from a single stock?* Or, on the other hand, *Have the different races of mankind, from the beginning of their existence, differed from one another in their physical, moral, and intellectual nature?*" These questions conflated the protocols of science with the received tradition of religion. For if every human descended from Adam and Eve, why were there now so many physical differences among them? What events and natural forces had caused them to diversify to such a striking degree? On the other hand, if human races had been created separately, as Agassiz and others claimed, did this nullify the Genesis account of human origins?

None of these speculations were wholly dispassionate. In America, ethnology would always be bound up with the question of slavery. As the *New York Times* put it, the new science could "hardly be dated beyond the present century" in America, but it had quickly become a flashpoint in the slavery debate, feeding the "volcanic fires smouldering below unconscious feet." Most American ethnologists believed that blacks had been separately created in Africa and endowed with lesser intellectual capacities than whites. (*Caucasian* was the preferred term for the latter, coined by a German ethnologist named Johann Friedrich Blumenbach in the mistaken belief that whites had originated near the Caucasus Mountains.) The differences presumed to exist between the races were copiously catalogued and illustrated in books that claimed the new authority of science.

Not everyone agreed with the separate creation model. Some scholars argued that humans were pliable, changeable, capable of extraordinary adaptations. They had sprung from a common ancestor and diversified over time. Advocates of this idea came to be known as *monogenists,* a term coined in 1857 to mean "advocates of a single origin." For the monogenists, the differences among races—in hair texture and skin color, for instance—were merely superficial variations. They did nothing to undermine the deep connection linking all people, a view that had the benefit of squaring with the traditional Christian belief that humanity had sprung from a single pair in the Garden of Eden. Monogenism received its first significant articulation by the British physician and ethnologist James Cowles Prichard, whose *Researches into the Physical History of Mankind* (1813) went through numerous editions over the first half of the nineteenth century and was eventually expanded into a five-volume treatise. According to Prichard, blacks and whites, Malays and Eskimos had all emerged from the same primal stock and were made in "the likeness of the Creator." In his opinion, "mankind constitute but one race or proceed from a single family." Physical differences were merely the products of "natural causes" such as climate, environment, and the availability of food "on a race originally uniform." Prichard believed the oldest human civilization was that of Egypt and therefore concluded that the human race had originated in Africa. In his most controversial statement, he argued that the original, "primitive stock of men were Negroes."

These ideas were eclipsed by the middle of the nineteenth century, however, by a handful of American ethnologists who came to be known as *polygenists*. Their version of human origins crystallized in the 1840s, when the Nashville physician Charles Caldwell composed his *Thoughts on the Original Unity of the Human Race.* Caldwell was the founder of the University of Louisville and an expert on fever and pestilential diseases. He also owned slaves. "We have no great objection to the theory which maintains, that each *species,* or distinctive and incommutable race of men, is the progeny of a single pair," he wrote. But he believed that whites and blacks were two separate and distinct species. Caldwell was equating *race* with *species*—a common error of the period—but it was a strategic mistake, allowing him to speculate that God had created exactly four categories of people: Caucasian, Mongolian, African, and American Indian. Each group was indelibly marked by distinct traits and attributes that implied an equally indelible and distinct hierarchy. One had only to examine an African male to see that his stomach was rounder, his blood darker, his penis larger—circumstances, Caldwell insisted, that placed him lower on the scale of human development and tended to "assimilate him . . . to the male ape."

At its most pernicious, polygenetic theory was an argument that removed people of African descent from the category of the human. Which isn't to say that monogenists were free from racism. Many who believed in a common origin for all people nevertheless assumed that blacks were a degenerate offshoot of the human family begun with Adam and Eve. But polygenists abandoned scriptural accounts entirely and depicted their conclusions as empirical fact. Samuel G. Morton, the Philadelphia physician who amassed a collection of some one thousand skulls from around the world and carefully measured and categorized them by race, claimed to prove the superiority of whites through the statistical analysis of cranial capacity. (His entire project was based on a false assumption. Variant brain size in humans actually has no correlation with intelligence.) The Mobile, Alabama, physician Josiah Nott and the English-born Egyptologist George Gliddon carried Morton's findings even further in their monumental *Types of Mankind,* a richly illustrated eight-hundred-page work published in 1854.

All these thinkers claimed to be advancing objectively verifiable knowledge, but they were not at times above appealing to the emotions; Nott asked readers to "Look first upon the Caucasian female, with her rose and lily skin, silky hair, Venus form, and well chiseled features—and then upon the African wench, with her black and odorous skin, woolly head and animal features." Typically, however, the racist conclusions espoused by polygenism were cloaked in the rhetoric of a neutral and disinterested science. As one ethnologist wrote, "It has been charged upon the views here advanced that they tend to the support of slavery. . . . Is that a fair objection to a philosophical investigation?" The author was Louis Agassiz.

No one understood the ramifications of Agassiz's polygenism better than Frederick Douglass. On a sweltering day in July 1854, Douglass gave a commencement address at Western Reserve College in Hudson, Ohio. This in itself was remarkable: black people in antebellum America were not invited to speak to graduating white students. Perhaps for this reason Douglass chose not to give one of his famous barn-burning abolitionist speeches (which was why the student body had invited him in the first place). He delivered instead a scholarly treatise entitled "The Claims of the Negro, Ethnologically Considered."

One of the era's greatest orators and a spellbinding storyteller, Douglass was easily the most famous black man in America. In a series of vivid and riveting autobiographies, he had described his hard-fought rise from slavery to freedom, suggesting that his story might one day be taken as a blueprint for black emancipation. Central to that narrative was his encounter, at age sixteen, with a "negro-breaker" named Edward Covey, who repeatedly beat and humiliated the young man until finally he "resolved to fight." For Douglass, this battle "was the turning point in my career as a slave." It structured his understanding of race relations in the United States, shaped his belief that slavery was a form of warfare enacted by one race upon another. Convinced that abolition could not be accomplished through moral suasion or legislative maneuvering alone, Douglass argued that life was struggle, that freedom had to be secured through resistance, conflict, and a fierce effort in which one's very being was risked but in which the slave might achieve "manhood," self-respect, and survival.

Accordingly, he began his commencement address in Ohio by describing the ideological struggle dividing the nation. "The relation subsisting between the white and black people of this country is the vital question of the age," he declared. "To the lawyer, the preacher, the politician, and to the man of letters, there is no neutral ground." Waving a copy of the *Richmond Examiner* before his audience, he read an excerpt from the Southern paper that exposed the moral depravity of those who supported slavery. According to the article, no slave could claim legal or moral rights— "BECAUSE HE IS NOT A MAN!"

There was nothing particularly new in this assertion. The supposed bestial and subhuman qualities of Africans had been used to justify slavery since its very beginnings. In 1787 the German historian Christoph Meiners was merely repeating an accepted commonplace when he declared that blacks had "no human, barely any animal feeling." What troubled Douglass, however, was the way pro-slavery activists now were using science to advance this claim. "It is remarkable—nay, it is strange," he said, "that there should arise a phalanx of learned men—speaking in the name of *science*—to forbid the magnificent reunion of mankind in one brotherhood."

Douglass had in mind people like Samuel G. Morton, Josiah Nott, and George Gliddon. But he was most concerned by Louis Agassiz, who seemed to believe God had fashioned whites for "higher things": for culture and Beethoven and classificatory schemata of the natural world. In his commencement address Douglass disputed "the Notts, the Gliddens [*sic*], the Agassiz, and Mortons," asserting that "what are technically called the negro race, are a part of the human family, and are descended from a common ancestry, with the rest of mankind." Humanity's shared inheritance not only reflected "most glory upon the wisdom, power, and goodness of the Author of all existence." It also fulfilled scripture. "The unity of the human race—the brotherhood of man—the reciprocal duties of all to each, and of each to all, are too plainly taught in the Bible to admit of cavil."

The problem, Douglass saw, was that the new "science" resisted moral arguments. It claimed to concern itself solely with facts and experience. Douglass argued that science was never pure, never entirely innocent of self-interested manipulation. Slavery was in truth "at the bottom of the whole controversy" over ethnology, and the debate between polygenists and monogenists was really a proxy battle "between the slaveholders on the *one* hand, and the abolitionists on the other." Indeed, the polygenists repeatedly disclosed their ideological commitments by distorting the very evidence they declared was neutral. Agassiz and the other scientists consistently "separate the negro race from every intelligent nation and tribe in Africa." (Douglass was thinking principally of Egypt.) And their ethnological manuals typically presented European faces "in harmony with the highest ideas of beauty," while the "negro, on the other hand, appears with features distorted, lips exaggerated, forehead depressed." It was clear that these so-called scientists had "staked out the ground beforehand, and that they have aimed to construct a theory in support of a foregone conclusion."

What troubled Douglass most about the polygenist theory was the way it provided scientific legitimacy to the malicious practice of equating black people with animals. The black abolitionist Henry Highland Garnet condemned the animalization of black people when he argued that slaveholders "endeavor to make you as much brutes as possible. When they have blinded the eyes of your mind—when they have embittered the sweet waters of the light which shines from the word of God—then, and not till then has American slavery done its perfect work." The prominent American abolitionist Theodore Wright Weld similarly observed, "The same terms are applied to slaves that are given to cattle. They are called 'stock'" and "female slaves that are mothers, are called 'breeders' till past child bearing." On the auction block, slaves were treated like cattle, pigs, horses: their muscles prodded, their mouths pried open, their sexual organs inspected before crowds of spectators.

The psychological toll of such treatment is almost impossible to imagine. Sarah Grimké, a white abolitionist writer raised among slaves in South Carolina, believed anyone who had suffered physical pain could conceive "the nakedness of some [slaves], the hungry yearnings of others, the wailing and wo[e], the bloody cut of the keen lash, and the frightful scream that rends the very skies." Much more difficult to comprehend was the effect of slavery upon one's personhood—one's *soul*. According to Grimké, the institution transformed a slave into "a thing, a chattel personal, a machine to be used to all intents and purposes for the benefit of another. . . . It would annihilate the individual worth and responsibility conferred upon many by his Creator."

Treating slaves like animals enabled slave owners to justify depriving their human property of the basic rights enshrined in the Declaration of Independence. While generalizations can be misleading—in 1844 a Kentucky judge declared, "A slave is *not* in the condition of a horse. . . . He is made after the image of the Creator. He has mental capacities, and an immortal principle in his nature"—the tendency to depict slaves as subhumans or as animals nevertheless became pervasive in the first half of the nineteenth century. It was bolstered by ethnologists who claimed that black people were a separate race or species, closer to animals, especially primates, and therefore possessing little in common with their white masters. No less an authority than Louis Agassiz enjoyed musing over the relationship between blacks and animals. "[Is it] not a little remarkable," he asked, "that the black orang occurs upon that continent which is inhabited by the black human race, whilst the brown orang inhabits those parts of Asia over which the chocolate-colored Malays have been developed[?]"

Douglass responded with thunderous anger. "Man is distinguished from *all* other animals," he exhorted his Ohio audience, "by the possession of certain definite faculties and powers, as well as by physical organization and proportions. He is the only two-handed animal on the earth—the only one that laughs, and nearly the only one that weeps." Human beings instinctively distinguished themselves from other animals, instinctively placed themselves above all other creatures. "Common sense itself is scarcely needed to detect the absence of manhood in a monkey, or to recognize its presence in a negro."

It is not exactly clear when Douglass first encountered Darwin's *Origin of Species* or whether he read it in any detail. He certainly became familiar with its arguments and would eventually adopt them in his attacks on scientific racism. In an 1864 lecture written at the height of the Civil War, he derided "a certain class of ethnologists and archeologists, more numerous in our country a few years ago than now and more numerous now than they ought to be and will be when slavery shall have no further need of them." The reference suggests how quickly Darwin's theory of common ancestry weakened the argument for the separate creation of human races. But if Douglass embraced Darwin's vision of common inheritance, he consistently evaded that portion of evolutionary theory that linked human beings to nonhuman species.

In 1854, as he concluded his commencement address at Western Reserve College, he indicated why this aspect of Darwinian theory would seem so odious. "Away, therefore, with all the scientific moonshine that would connect men with monkeys," he told the Ohio graduates, "that would have the world believe that humanity, instead of resting on its own characteristic pedestal—gloriously independent—is a sort of sliding scale, making one extreme brother to the ourang-ou-tang, and the other to angels, and all the rest intermediates!"

10

The What-Is-It?

From his stuffed, rococo office on Manhattan's bustling Broadway, Phineas Taylor Barnum eyed Paul du Chaillu's exhibition with growing concern. The "Prince of Humbugs" (a moniker he cheerfully invented for himself) had courted controversy since the 1830s, and he did so for a very simple reason: controversy sold. As a twelve-year-old clerk in his father's Connecticut business, P. T. Barnum learned the invaluable lesson that "even in a country store . . . exceptions to the general rule of honesty" were frequent, because "nearly every thing was different from what it was represented." Barnum's first foray into false advertising involved a former slave named Joice Heth. In 1835 he made fifteen hundred dollars a week displaying her, claiming she was 161 years old and had served as a wet nurse to George Washington. When she soon died, he charged fifty cents per person to view her autopsy.

Barnum organized dog shows, flower shows, fat baby contests. He brought the soprano Jenny Lind to America on a wildly popular singing tour, and he traveled to Europe with a dwarf he christened General Tom Thumb. In 1842 he opened his American Museum on the corner of Broadway and Ann Street in lower Manhattan. There he displayed his promiscuous and ever-growing collection of natural and human artifacts. ("My organ of acquisitiveness must be large," Barnum admitted in his autobiography, one of the best-selling books of the nineteenth century, "or else my parents commenced its cultivation at an early period.") The museum contained picture galleries, freak shows, and a full-fledged zoo. There was a preserved elephant, a sword from Oliver Cromwell's army, a mummy from Thebes, and a severed arm supposedly belonging to a pirate. Barnum draped an enormous American flag across the five-story building and bedizened the rest in gaudy advertisements. He bathed the whole affair in limelight, and he hired the worst musicians he could find to sit on the balconies and play execrable music. His theory was that people outside would hurry inside to avoid the assault on their ears.

He was a blowhard, a genius at self-advertisement, a con artist nonpareil—but he was also one of the nation's premier collectors of plants and animals and assiduously kept up with the science of his day. When he died in 1882, Smithsonian curator Spencer Fullerton Baird asked Barnum's family for a copy of the great showman's death mask to include in the institution's "series of representations of men who have distinguished themselves . . . as promoters of the natural sciences." Barnum employed dozens of collecting agencies to ship crates of natural specimens to his museum. Interspersed with sensationalist artifacts, Barnum's natural history collection was so comprehensive that even a skeptical Henry David Thoreau visited the museum three times.

Bad investments bankrupted Barnum in the 1850s. (Emerson claimed the showman's financial failure made "the gods visible again.") But he was back in business by 1860, and as luck would have it, there were plenty of controversies to exploit. The *New-York Tribune* ran a spoof entitled "A Fortune for Barnum" about a most unusual freak: "A man has been found who has never heard of John Brown!" Barnum soon exhibited a full-size wax figure of the abolitionist raider, along with a letter he had written and a knife said to belong to his son. He also staged a version of Dion Boucicault's wildly popular melodrama *The Octoroon, or Life in Louisiana,* which managed to offend pro- and antislavery audiences alike. Some newspapers called the play "John Brown on Stage," and the pro-slavery *New York*

Herald announced that "we wage war against the 'Octoroon,' and declare, in the name of the good citizens of the metropolis, that neither that nor any other play of the same character should be performed." Barnum made money hand over fist.

But recently he had been distressed to learn that a diminutive foreigner had had the temerity to exhibit a handful of African oddities so near to his own palace of amusement. Du Chaillu's rented space was attracting small crowds that gathered to marvel at the skins and skeletons of the so-called "man-monkey" and to listen to the adventurer's account of killing a male gorilla. "With a groan that had something terribly human in it and yet was full of brutishness," du Chaillu said, "he fell forward on his face. The body shook convulsively for a few minutes, the limbs moved about in a struggling way, and then all was quiet—death had done its work, and I had leisure to examine the huge body."

Within weeks Barnum launched a counterattack. In early January 1860 posters appeared along Broadway advertising the "MOST MARVELOUS CREATURE LIVING." In slightly smaller print, a pair of tantalizing questions: "Is it a lower order of man? Or is it a higher development of the Monkey?"

Barnum had long been obsessed with metamorphosis and transformation. The fascination stemmed in part from his meteoric rise to prominence as the wealthiest showman in the nation. But his interest in creatures that defied neat categories of classification also reflected the topsy-turvy dynamics of a raw and boisterous democratic nation, a place where class and racial boundaries were fluid and unstable. One of Barnum's popular hoaxes was the so-called "Feejee Mermaid," a baboon head grafted to the tail of a cod. The oddity drew hundreds of thousands of spectators. In the 1840s he first began displaying something he called the What-Is-It?—a creature that exhibited traits of at least two distinct species. The first of these was a live orangutan. In posters Barnum described the primate as the *"Connecting link between Man and Brute!!!!"*

"Advertising," Barnum liked to say, "is to the genuine article what manure is to the land,—it largely increases the product." Therefore early in 1860 the city's newspapers began to print a story about a new specimen at the American Museum that had been discovered in Africa by a band of explorers in search of gorillas. Coming upon a small group of these animals, which lived in the canopy of the jungle, the explorers pursued them. Two escaped; two more died on their way to America. Barnum just happened to own the sole survivor. "He possesses the skull, limbs, and GENERAL ANATOMY OF THE ORANG-OUTANG," he now proclaimed, "with the actual COUNTENANCE OF A HUMAN BEING!"

Barnum's latest What-Is-It? was an instant hit. Visitors to his museum crowded into a room and watched as the marvel stood in a cage and made feeble, goatish cries. Posters claimed the What-Is-It? possessed tremendous strength. "There is apparently more strength in his hands and arms than in all the rest of his body combined." Other peculiarities were detailed. "His legs are crooked, like those of the Orang Outang. . . . The WHAT IS IT'S FOOT is narrow, slim, and flat, and has a long heel like that of the Native African. The large toe is more like a man's thumb." Here, apparently, was the long-awaited missing link that connected primate to human. Or as one spectator commented, "his anatomical features are fearfully simian, and he's a great fact for Darwin."

That spectator was George Templeton Strong, the Manhattan lawyer, diarist, and acquaintance of Charles Loring Brace, who had just finished reading *On the Origin of Species* when he decided to stop by Barnum's "to see the much advertised nondescript, the 'What-is-it.'" He was still pondering Darwin's book, which he had read carefully, perusing some portions twice. The more he ruminated on the work, the more he found himself annoyed by its naturalism: its insistence on a mechanistic explanation for all of nature's phenomena. The author refused, so far as Strong could tell, to rely on any supernatural explanations—on *God*—for the origin of life. "Darwin cannot understand," Strong wrote in his diary, "why Omnipotent Power and Wisdom should have created so many thousand various types of organic life. . . . He wants to shew that the original creative act was on the smallest scale and produced only some one organism of the humblest rank but capable of development into the fauna and flora of the earth, from moss to oak and from monad to man, under his law of progress and natural selection."

Strong possessed a ruthlessly logical temperament (he was a lawyer, after all), and he considered this a case of interpretive overdetermination. Darwin had become too enamored with his theory and therefore refused to consider other possibilities. He "has got hold of *a* truth," Strong wrote, "which he wants to make out to be *the* one generative law of organic life." In Strong's opinion, miraculous production constantly occurred in nature. For instance, "certain elemental atoms of lime, silex, carbon, oxygen, hydrogen, and so forth are daily and hourly 'commanded to flash into' organic wood fibre, cellulose . . . and so on." (He was mistaken: the growth of plants did not require supernatural explanation, as his list of elements had suggested.) But Darwin demanded other concessions. First was the inordinate amount of time—"thousands of millions of millions of years"—required for his natural process to work. Then there was the comparative silence of the fossil record. If "flying fish by successive minute steps of progress through countless ages become albatrosses, and flying squirrels bats," Strong asked, why were there no records in the stratified layers of geology? Where were the missing links?

Ultimately, Strong's greatest objection to Darwin's theory was personal. He disliked the implication that "man is the descendant of some ancestral archaic fish." Diminishing humankind's importance in this way ultimately suggested the fragility of civilization. It revealed the flimsy consensus that propped up morality. It led to anarchy. In Strong's mind, human society was chaotic enough, New York already like some huge aquarial garden filled with sharks and fish, each trying to consume smaller and more vulnerable species in order to survive and propagate. Without the firm and steady hand of a ruling elite, Strong worried, the entire city threatened to disintegrate into a riotous free-for-all.

He entered the tumult of Barnum's museum, where crowds of men and women jostled in the great hall. Strong was greeted by a pair of "White Negroes": albinos with fair hair and African features who stood in the entry room to be examined, allowing visitors to step as close as they wanted to them. But the majority of people were bunched together before a cage containing the What-Is-It?

Strong was unimpressed. "Some say it's an advanced chimpanzee, others that it's a cross between nigger and baboon," he wrote in his diary that evening. "The showman's story of its capture . . . by a party in pursuit of the gorilla on the western coast of Africa is probably bosh." Still, there was something strange and uncanny about the creature. Standing inside a narrow cage, his hands grasping its bars as he stared back at the crowd, the What-Is-It? mewed and bellowed and made unintelligible noises while the crowd stood in expectant silence.

Strong returned to Barnum's museum the next day, this time bringing his friend and chess partner, George Christian Anthon. The two friends stood for a long time inspecting the What-Is-It? The creature was dressed in a fur smock, and there was something misshapen about his head, but instead of growling ferociously he seemed playful, even childlike. The two friends agreed that Barnum's establishment contained other animals that were much more interesting: "a grand grizzly bear from California, a big sea lion, a very intelligent and attractive marbled or mottled seal (*phoca vitulina?*), a pair of sociable kangaroos." The What-Is-It?, Strong decided, was on the other hand, "palpably a little nigger and not a good-looking one."

He was not an *it* at all, of course, but an eighteen-year-old named William Henry Johnson, the son of two former slaves so impoverished and without prospects that they had allowed Barnum to exhibit their sixth child in exchange for a dollar a day. Historians have long assumed that Johnson was born with microcephaly, a neurological disorder that results in an unusually small skull; Barnum accentuated this by shaving Johnson's head and leaving only a topknot. But those afflicted with microcephaly typically experience diminished brain capacity and shortened life expectancy. The evidence is somewhat mixed about his mental life, but Johnson performed as one of Barnum's freaks for the next sixty years, eventually appearing before the public as "Zip," a nickname he retained until his death in 1926. Humiliating as his occupation was, he was well compensated. Barnum soon increased his salary to one hundred dollars a week, an enormous sum for the period, and upon his death Johnson owned a number of houses and had saved a substantial sum of money. Like Barnum, he had risen from modest and obscure circumstances to wealth and notoriety.

What strikes a modern viewer is just how *normal* Johnson appears in photographs. Far from the grotesque creature featured in Barnum's posters or in Currier and Ives's popular lithographs, Johnson manages to convey a quiet dignity even while clothed in Barnum's ridiculous costumes. One story, almost certainly apocryphal, has it that upon his deathbed Johnson whispered to his sister, "Well, we fooled 'em for a long time." But one wonders how many people actually *were* fooled. P. T. Barnum was a genius at turning the world upside down, of subverting norms and eroding barriers. But his brilliance lay in eventually restoring those norms. He became immensely wealthy not by unsettling audiences but by returning them to the safety of their preconceptions. George Templeton Strong's belief in the separation between humans and animals may have been threatened by Johnson's performance, but it was reestablished as soon as he recognized the hoax. Whether audiences believed that the What-Is-It? was really a missing link or simply a disabled black man, their own self-image was immeasurably enhanced. As usual, Barnum had devised the perfect attraction.

Charles Loring Brace walked past the American Museum nearly every day. He would have taken note of the What-Is-It? if for no other reason than that posters advertising the attraction were plastered up and down Broadway. But he was also fascinated by ethnology and would have been alert to the way in which Barnum's creature fit—or failed to fit—into ethnological categories. After all, he was writing a manual about the controversial science.

The book was for the general reader, and it drew upon Brace's preoccupation with the way climate and geography "modified or changed the bodily type of a race." For nearly a decade he had devoted his leisure hours to studying the new science, drawn to historical accounts of people who had emigrated from one place to another, forming his own hypotheses about how such migrations altered human characteristics. In 1854, he devoured Johann Georg Müller's *Geschichte der Amerikanischen Urreligionen,* a book on Native American religion. (He was interested in how nomadic living shaped a people's spiritual organization.) A year later he wrote the abolitionist minister Theodore Parker to say, "I am anxious to talk over some ethnographic matters with you," adding that he had already immersed himself in the historical sagas of Scandinavia and hoped to begin studying Eastern religions soon.

But it was Darwin's book that suddenly energized the project. The *Origin* revealed to Brace how polygenism might be combated, which is to say, it provided him with a new way to attack slavery. "It is the deepest feeling of my heart that no darker stain rests on this country than this slavery," he had observed as a young man. Now, in 1860, he complained that "throughout Europe, American science . . . has become identical with perverted argument for the oppression of the negro." The politics of slavery had polluted scientific inquiry on both sides of the Atlantic, Brace claimed, echoing in many ways the argument Frederick Douglass had made six years earlier. Southern slave interests had corrupted free and dispassionate investigation. "The shadow of our national sin has fallen even on the domain of our science, and obscured its noble features to the world." Urgently needed was a book that would correct this problem by returning the study of human societies to its "solid basis of facts and inductive reasoning."

What George Templeton Strong claimed about Darwin was in fact true about Brace: he had "got hold of *a* truth which he wants to make out to be *the* one generative law of organic life." From the moment Brace first read Asa Gray's copy of the *Origin,* he believed Darwin's theory could be extended into virtually every aspect of human life. For instance, natural selection proved to him the inevitability of human salvation. The argument went like this: "Evil seems to me destructive,—good preservative." This meant that innate virtue—an inborn sense of right and wrong—was an adaptive advantage. It preserved the human race from degradation, leading it ever upward toward its evolutionary destiny of total perfection. Put another way, morality was influenced by evolutionary laws. "The idea of this age," Brace wrote his cousin Henry Ward Beecher, the reformist minister and the brother of

Harriet Beecher Stowe, "is of slow growth, especially of all moral things." Social reformers and clergymen might labor to effect sudden conversions in their constituents, but the transformation of the soul was in fact "influenced by ten thousand imperceptible causes, and [a person's] salvation is the slowest of all things." For Brace, Darwin's theory enabled one to think about any subject involving growth and progress: history, politics, race, moral development. "It is remarkable how the application of the law of natural selection is influencing now every department of scientific investigation," he wrote. "I think it furnishes what historians and philosophers have so long sought for, a law of progress, and Darwin states the glorious point to which mankind shall eventually advance."

Brace was not alone in reading the *Origin* as a commentary on American ethnology. "We all know how energetic, I might almost say how angry, a war has just been waged for and against the doctrine of the unity of the human race," declared the Washington *Daily National Intelligencer* in a February review of Darwin's book. "But here [with the *Origin*] we have a bolder fight on the affirmative side. Here it is not the unity of one allied family, such as man confessedly is, but the unity of creation, the origin of all species from one central unique source." Reviewers for the American popular press consistently understood Darwin as having provided a theory that showed that black and white people were related. Antislavery newspapers praised the new book for its implicit attack on the popular ideas of Louis Agassiz and other ethnologists.

In truth, Darwin had refrained from addressing this issue in the *Origin,* in part because he was unwilling to claim more for his theory than it could adequately answer. But he also assumed that readers would draw their own conclusions and make their own associative leaps. He had been raised in an antislavery family, his grandfather, Josiah Wedgwood, having produced the famous jasperware cameo of a chained slave on his knees with the slogan, *Am I Not a Man and a Brother?* His extended family had contributed enormous sums to help end the British slave trade. Darwin had witnessed slavery firsthand. During his voyage aboard the *Beagle,* he heard crews of slaves singing their morning hymns as they toiled beneath a slashing Brazilian sun. He encountered one slave-owning woman in Rio de Janeiro who "kept screws to crush the fingers of her female slaves." Children no taller than his waist, barefoot and often naked, were struck with horsewhips and hurled to the ground.

These slaves had been transported to Brazil in ships no larger than the *Beagle,* packed and chained in dank filthy holds that accommodated as many as five hundred men, women, and children. Disease killed as many as half the captives. Some chose to leap into the Atlantic rather than endure life as a slave. Darwin listened to Brazilian slave owners defend the cruel practice by calling black people "the vilest of the human kind" and comparing them to brute creatures. African women were "libidinous and shameless as monkeys," they said, and the entire race were so close to primates that "an oran-outang husband would [not] be any dishonour to a Hottentot female."

Riding a ferry one day, Darwin tried to speak to a slave who understood very little English. The man was huge, powerfully built, unresponsive. Darwin raised his voice. He repeated himself and spoke slowly—all without result. Once, as he tried to convey his question by gesturing, his hand fleetingly passed within an inch of the slave's face. "He, I suppose, thought I was in a passion, and was going to strike him," Darwin recalled, "for instantly, with a frightened look and half-shut eyes, he dropped his hands. I shall never forget my feelings of surprise, disgust, and shame, at seeing a great powerful man afraid even to ward off a blow, directed, as he thought, at his face. This man had been trained to a degradation lower than the slavery of the most helpless animal."

On yet another occasion, as he strolled through the sultry stillness of a remote village near the mangrove swamps of Pernambuco, Darwin "heard the most pitiable moan." Instantly he realized "some poor slave was being tortured." A wave of horror washed over him: "I was as powerless as a child even to remonstrate." The moment haunted him for the rest of his life. "To this day, if I hear a distant scream, it recalls with painful vividness my feelings." By the time the *Beagle* completed her voyage around the world, Darwin had seen enough. "I thank God," he wrote, "I shall never again visit a slave-country."

Brace would become the first person to write an ethnological interpretation of Darwin. In 1860 the New York philanthropist wrote letters to prominent scientists, rifled through New York's private libraries, and took copious notes on previous works of ethnology. He compared monogenist and polygenist theories and delighted in the fact that Darwinism made scriptural *and* scientific sense. He studied craniometry, read travel narratives, perused the latest issues of the *American Journal of Science*. His manual on ethnology grew in length and complexity; as is so often the case with such work, the more he studied, the more he realized how much he still needed to know.

Which presented a problem. Like his acquaintance Henry David Thoreau, he found himself increasingly lost in the thickets of his research, immersed in a welter of facts and reports, statistics and accounts. Brace had divided his study into loose groupings: Africa, Europe, Asia, the Pacific Islands. But these categorizations did little to simplify the prodigious amounts of information he had acquired. After all, how was one to think of the inhabitants of northern Africa—as black, brown, or quasi-European? How was one to think of peoples who had migrated thousands of miles, mixing with native populations and becoming something new? Increasingly Brace found it difficult to decide on a single direction for his book.

Fortunately—and at just the right moment—Asa Gray provided the clue. This was in his first essay about Darwin's book for the *Atlantic*.

11

A Spirited Conflict

Years later, when the reclusive poet Emily Dickinson archly wrote that "we thought Darwin had thrown 'the Redeemer' away," she was almost certainly remembering Asa Gray's series of articles about the *Origin of Species* that appeared in the *Atlantic Monthly* during the summer of 1860. Dickinson's father never purchased Darwin's book, but he did subscribe to the *Atlantic,* and his daughter assiduously scoured the magazine each month. (It was the prominent *Atlantic* contributor and abolitionist Thomas Wentworth Higginson to whom she wrote in 1862, asking whether her verses "breathed" and initiating a lifelong correspondence.) As with many Americans, Dickinson's first encounter with Darwin was through Gray's reviews.

By February 1860 the book had already received notices in the *New York Times,* the *Boston Daily Advertiser,* and the *Springfield Republican,* as well as in *Harper's New Monthly Magazine,* the *North American Review,* the *New England Review,* the *Christian Intelligencer,* the *Methodist Quarterly Review,* the *American Theological Review,* and the *Biblical Repertory and Princeton Review.* Some of these publications focused on the work's ethnological implications. Others criticized the way Darwin's ideas undermined religion. One pious reviewer noted that natural selection stripped divine intention from the center of the physical universe and replaced it with chance and accident. The *Origin* was nothing less than a "sneer at the idea of any manifestation of design in the material universe," and its theories "repudiate the whole doctrine of final causes," rendering obsolete "all indication of design or purpose in the organic world." Another indignant critic, this one for the *Christian Examiner,* claimed that Darwin's central idea was "neither more nor less than a formal denial of any agency beyond that of a blind chance."

Gray would attempt to counter these attacks in his three-part series for the *Atlantic,* but before he could do so, he had to complete a lengthy review for the *American Journal of Science,* which would be published in March 1860. There he portrayed Darwin's theory as a young combatant in an age-old war of ideas. Natural selection could be considered "atheistical," he admitted, because it did not require God for its operation. But this was true of *all* theories that tried to explain physical phenomena. Darwin's ideas were no different from those of Isaac Newton, whose "theory of gravitation and . . . nebular hypothesis assume a *universal and ultimate* physical cause, from which the effects in nature must necessarily have resulted." No one seriously believed that gravity banished God from the universe. Gray wondered how any "scientific man" could believe that a material connection between diverse plant and animal species "is inconsistent with the idea of their being intellectually connected with one another through the Deity"?

The "scientific man" he referred to was Agassiz. Throughout his March article Gray argued for cautious, inductive reasoning—something that was impossible when one routinely made a priori assumptions about the role of a divine Creator in the workings of nature. For Gray, there were two ways of understanding the natural world: the transcendental and the material. Both approaches offered appealing explanations for how the universe had come into being, but while the idealistic view assumed an "independent, specific creation of each kind of plant and animal in a primitive stock," Darwin's more naturalistic description of creation began instead with "a single pair, or a single individual" that gradually transmuted into separate species. Gray admitted that the developmental theory, as it was commonly referred to, broke down a centuries-old belief that species were stable and immutable. But this belief made no sense when one examined the evidence. *All* species varied to some degree, some quite remarkably so. Offspring were never simply copies of one another. The source of these variations remained unknown, but their occurrence implied the possibility of change and differentiation.

The main problem with the idealist view of species was not its misreading of evidence, Gray asserted. It was that it relied entirely on a supernatural explanation. Agassiz's science was "theistic to excess." This was not to say God was absent from the creation of life—quite the opposite, Gray believed. But science was a method concerned with observing organisms, processes, and laws in the *material* world. There was simply no place in scientific inquiry for leaps of faith or speculation about the unseen. By referring "the phenomena of both origin and distribution [of species] directly to the Divine will," Gray wrote, the idealist theory removed the study of organic life from "the domain of inductive science."

In sharp distinction, Darwin's new theory was grounded upon close and careful observation. It was based upon data that could be shared and verified. And it revealed something everyone intuitively knew because everyone had observed or experienced it: that "plants and animals are subject from their birth to *physical* influences, to which they have to accommodate themselves as they can."

Gray admired Darwin's book because it reinforced his conviction that inductive reasoning was the proper approach to science. Darwin's book was both a primer and a meta-text about scientific theories: about *their* struggle to survive. Hypotheses and conclusions thrived only when they adapted to the harsh demands of evidence and repeatable testing. "A spirited conflict among opinions of every grade must ensue," Gray observed about the theories of natural selection and special creation, "which—to borrow an illustration from the doctrine of the book before us—may be likened to the conflict in Nature among races in the struggle for life."

This now seems like a rather obvious point to make. At the time it was radical. Gray was suggesting that whether or not Darwin's ideas were correct was ultimately less important than how those ideas were to be evaluated. To reach a conclusion about evolutionary theory by way of emotion or inherited belief or any criteria other than empirical reasoning was intellectually untenable. It was dishonest and unsound. It was *unscientific.* What mattered for Gray was the careful judging and clear-eyed balancing of data. In order to grapple with Darwin's ideas, one had to follow the evidence wherever it led, ignoring prior convictions and certainties or the narrative one *wanted* that evidence to confirm. Although he did not say it in quite these terms, Gray was suggesting that readers of Darwin's book had to be open to the possibility that everything they had taken for granted was in fact incorrect.

From his perch as the most eminent scientist in America, Louis Agassiz learned of Gray's review with dismay. Agassiz had received an author's copy of the *Origin* in December, around the same time as Gray. Darwin had written to him, "As the conclusions at which I have arrived on several points differ so widely from yours, I have thought (should you at any time read my volume) that you might think that I had sent it to you out of a spirit of defiance or bravado; but I assure you that I act under a wholly different frame of mind. I hope that you will at least give me credit, however erroneous you may think my conclusion, for having earnestly endeavored to arrive at the truth."

Agassiz never replied.

He did, however, read the book—or at least a portion of it. Like Gray, he read with a pencil in hand, making notes as he went along. But those notes were of a decidedly different cast. "A sentence likely to mislead!" he scribbled. "This is truly monstrous." About a quarter of the way through the *Origin,* he indignantly summed up what he took to be the author's flawed methodology. "The mistake of Darwin has been to study the origin of species among domesticated animals instead of wild ones; his results concerning species are founded not on an investigation of species themselves but on an investigation of breeds." He wasn't finished. A few pages later Darwin indirectly attacked Agassiz by discussing the remarkable similarity of species in the same group or family. "On the view that each species has been independently created," Darwin wrote, "I can see no explanation of this great fact in the classification of all organic beings." Agassiz replied with exasperation in the margin: "Why not? does the excellences of the classification make it less likely to be the results of intelligent Creation? more likely to be the result of physical creation?" And when Darwin asserted that "every organic being is constantly endeavoring to increase its numbers," Agassiz sputtered, "Why sir, there is room enough for all the ducks of the world in the forests of North America." He forced himself to read a few more pages and then, in exasperation, skipped the second half of the book and hastily read its conclusion.

In January 1860 Agassiz discussed the book in public for the first time. "What has the whale in the arctic regions to do with the lion or the tiger in the tropical Indies?" he asked a large audience at the Boston Mercantile Library Association. "There is no possible connection between them." Agassiz was nettled by Darwin's hypothesis that all creatures were biologically linked. To him it was inconceivable that whales and lions could be linked by anything other than their mutual conception in the mind of the Creator. "There is behind them & anterior to their existence, a thought," Agassiz thundered. "There is a design according to which they were built, which must have been conceived before they were called into existence." He then fell back to his customary theme: "Whenever we study the general relations of animals, we study more than the affinities of beasts. We study the manner in which it has pleased the Creator to express his thoughts in living realities; and that is the value of that study for intellectual Man."

The problem with evoking the Creator to explain the natural world, of course, was that it foreclosed other hypotheses. It favored dogma at the expense of logic and empiricism. "Tell Darwin that Agassiz has *again failed* to provide his promised criticism on Darwin for [the] Jour[nal]," Gray wrote Hooker shortly after the speech at the Mercantile Library Association. "I do not wonder that he hesitates to commit himself to print. I really think his mind has deteriorated within a few years."

Unlike Agassiz, Gray returned to Darwin's book again and again, rereading passages, marking fresh selections, savoring favorite paragraphs. The *Origin* was stuffed with one delicious detail after another. The tendency to catch rats instead of mice was a heritable characteristic of cats. A ripe asparagus plant would float in salt water for twenty-three days. All this minutiae added up to something; it formed a larger pattern, a coherent argument. One of the book's signal accomplishments was to show how even the most trivial interaction between organism and environment was rich with meaning. Such tiny relationships, such insignificant causes and effects, could be decoded; they were in fact the very warp and woof of nature, diverse threads woven together to create a beautifully complex tapestry of life.

This perspective recommended itself to Gray because it revealed how experience and observation together produced facts. Like words in a sentence or pigments in a painting, facts were the essential building blocks of science, leading to the construction of plausible hypotheses, to sound conclusions. The cabinets of Gray's herbarium—stuffed with tens of thousands of dried plants, each neatly labeled, each examined beneath his microscope and studied for affinities and slight variations—were the repositories of facts. Gray believed that a person could not begin to say anything sensible about how plants grew or thrived, how they were distributed across the Earth or how they interacted with one another in a given environment, until that person had mastered as many of these facts as humanly possible.

Gray's *American Journal of Science* article occasionally disagreed with Darwin. It raised questions about reversion, or the tendency of offspring to exhibit characteristics of older ancestors. And Gray worried, like George Templeton Strong and others, about the absence of fossil records. He quibbled over Darwin's phraseology. But just below the surface of these concerns, he repeatedly challenged Agassiz's notions about science, argumentation, and the production of hypotheses. And he took Agassiz to task on the matter of race.

It had long been an open secret at Harvard that Agassiz felt a visceral repulsion toward black people. His Swiss upbringing might have conditioned this attitude, which would have been reinforced when he studied with Cuvier, who once described Africans as "the most degraded of human races, whose form approaches that of the beast and whose intelligence is nowhere great enough to arrive at regular government." But there was something particularly violent about Agassiz's reaction to the black servants he encountered for the first time during his tour of America. Traveling with Gray along the eastern seaboard in December 1846, he wrote a long letter to his mother describing the American scene. "It was in Philadelphia that I first found myself in contact with Negroes," Agassiz described, proceeding to detail his disgust at the physical characteristics of black hotel servants and waiters. The "feeling that they inspired in me," Agassiz confessed, "is contrary to all our ideas about the confraternity of human types (genre) and the unique origin of our species. . . . What unhappiness for the white race—to have tied their existence so closely with that of Negroes in certain countries! God preserve us from such a contact."

Agassiz did not condone slavery. He thought the institution in appalling taste, cruel to its victims, demeaning to its practitioners. Over time he even came to consider himself a reluctant abolitionist, advocating, when pressed, for the legal equality of blacks. But he never swerved from his basic conviction, forged in the moment when a black man waited on him, that people of African descent had been created separately from Europeans. A decade before reading Darwin's book, in 1850, Agassiz traveled to South Carolina to address a group of Southern scientists who were particularly interested in his ideas about the separate creation of races. While there he asked his host, the physician Robert W. Gibbes, if he might examine slaves from various African tribes; he wished to catalog the anatomical differences that separated them from whites. Half a dozen slaves were procured from nearby plantations and brought to Agassiz. They stood before him, neither smiling nor looking directly at him, and tried to answer his questions. At some point he asked each one, male and female, to strip to the waist, so that he might study them better. "Agassiz was delighted with his examination of Ebo, Foulah, Gullah, Guinea, Coromantee, Mandingo, and Congo negroes," Gibbes later recalled. The great naturalist had apparently seen enough to confirm him in his belief that "they have differences from other races."

Gibbes may have overstated the case somewhat. Agassiz never went as far as some polygenists; he never claimed that blacks and whites belonged to separate species. But he did believe they had been created mutually exclusive from each other, fashioned by God to thrive in their separate provinces. For this reason he was especially horrified by the prospect of the two races intermingling. His principal opposition to slavery was not that it converted people into chattel property or oppressed unfortunate humans by coercing them to perform backbreaking labor. Rather, he was horrified that the institution fostered interbreeding—what was then known as "amalgamation." The prolific intermixing of blacks and whites through cohabitation and sexual congress was for Agassiz an abomination. It blurred and confounded racial categories. It violated the calm and logical order of the universe and disregarded nature's most fundamental divisions, which is to say, it violated God's will.

"Conceive for a moment the difference it would make in future ages," Agassiz wrote, "for the prospect of republican institutions and our civilization generally, if instead of the manly population descended from cognate nations [he meant Europe] the United States should hereafter be inhabited by the effeminate progeny of mixed races, half Indian, half negro, sprinkled with white blood." This alarming future was not only possible but in fact probable, given the close proximity of two races that were "more widely different from one another than all the other races." It was simply inconceivable that blacks and whites could live together as equals. Amalgamation, he added, was "one of the most difficult problems upon the solution of which the welfare of our own race may in measure depend."

Now, in his article for the *American Journal of Science*, Asa Gray waded into the polygenesis controversy. He began with an obvious scientific point: according to Linnaeus, two species could not interbreed or produce fertile offspring. That blacks and whites quite clearly *did* produce offspring together had been dismissed by polygenists, who claimed that biracial children were infertile. "The best marked human races might offer the most likely case" for testing these assumptions, Gray mused. "If mulattoes are sterile or tend to sterility, as some naturalists confidently assert," that would afford evidence of two distinct human species. "If, as others think, no such tendency is made out, the required evidence is wanting." He hypothesized that human races might *one day* become separate species, given enough time and separation from one another. The crucial point, however, was that they shared the same inheritance—a claim his esteemed colleague, Louis Agassiz, disputed without providing evidence.

12

Into the Vortex

Crossing the bridge over the frozen Musketaquid River, where he noticed Emerson's children skating with friends, and then walking along the frozen mud-churned streets of Concord, Amos Bronson Alcott talked. He talked to women carrying parcels beneath the snow-covered awnings. He talked to children on their way home from Franklin Sanborn's school. He talked to farmers and carpenters, hot and red-faced despite the cold. Earlier in the day he had invited his friend Henry David Thoreau into his parlor, where the two men warmed themselves before the fire and gazed at the marble bust of Plato perched atop his crammed bookcase. As usual, he had talked.

On this particular day in February 1860, however, Bronson Alcott was not speaking about the divinity within, or about a new heaven on Earth, or even about the pressing need for a decent geography book for the pupils of Concord, his current pet project. He was instead consumed with the accomplishments of his second-oldest daughter, Louisa, whose melodramatic short story "Love and Self-Love" had just been published by the *Atlantic Monthly*. On this blustery day he carried a copy of the most distinguished literary journal in the country and was proudly displaying its table of contents to everyone.

Louisa was twenty-seven that year. Wry, ironic, alternately ebullient and stormy, she longed to become a best-selling author, as famous and consequential as Harriet Beecher Stowe, as enduring as her idol Charlotte Brontë. Like those authors, she was a cultural sponge; she absorbed her society's concerns and contradictions, its fears and aspirations, and she transformed these things into plot, motive, resolution. Perhaps because she created characters who represented various points of view, she was more accepting of new ideas than her father. Bronson had settled upon his philosophy decades earlier, and it remained a formidable support, a bulwark against the changing world. Louisa was more open to the Darwinian ideas then coursing through Concord's intellectual circles, so much so that she quickly absorbed them into the apprentice fiction she wrote that year.

She had long dreamed of leaving the provincial confines of Concord, of departing the old brown house she called "Apple Slump," of visiting the pyramids in Egypt and the boulevards of Paris. She raged at the limitations imposed on womanhood in the nineteenth century, toying with the idea of becoming a stage actress, then a nurse. Mostly she wanted to write—to direct her prodigious energies into a dream world where candid, fearless young women fulfilled their duty and achieved love, all without compromising their inner selves. In *Little Women,* the work that was based upon life at Orchard House and that eventually fulfilled her aspirations for money and fame, Alcott provided a vivid portrait of her writing life. Speaking of her alter ego, Jo March, she wrote: "Every few weeks she would shut herself up in her room, put on her scribbling suit, and 'fall into a vortex,' as she expressed it, writing away . . . with all her heart and soul, for till that was finished she could find no peace." There was something transcendent, almost ecstatic, about the experience. "When the writing fit came on, she gave herself up to it with entire abandon, and led a blissful life, unconscious of want, care, or bad weather, while she sat safe and happy in an imaginary world, full of friends almost as real and dear to her as any in the flesh." Louisa ironically deployed her father's language when she noted that the "divine afflatus usually lasted a week or two." Eventually she emerged from her "'vortex' hungry, sleepy, cross, or despondent."

Her energy found other outlets. She went outdoors, bunched her dress in one hand, and did what practically no other young woman of her period did: ran. She sped down country lanes and across pastures, sometimes through town, her dark hair streaming loose from its ribbons, her face flushed with exertion. And she proselytized on behalf of abolition. Like her mother, she belonged to the Concord Ladies' Anti-Slavery Society, attended lectures that condemned Southern slavers, distributed pamphlets, and collected funds. John Brown's visit to the village the year before had electrified her, and she had reacted to the news of his capture and arrest with outrage.

Mostly, though, she wrote. Brown's death wrung a poem from her that conveyed her reverence for the executed man:

> *There blossomed forth a grander flower in the wilderness of wrong,*
> *Untouched by Slavery's bitter frost, A soul devout and strong . . .*
> *No monument of quarried stone, No eloquence of speech,*
> *Can 'grave the lessons on the land His martyrdom will teach.*

She would always remember 1860 as a particularly "happy year," in part because it heralded her entrance into professional authorship. But she was also referring to her hard-won emergence from another sort of vortex—a profound depression that had nearly swallowed her up in 1858 when her sister Lizzie died from scarlet fever. Two weeks later her older sister Anna was betrothed, and Louisa felt abandoned, deprived of her two most important confidantes. Weighed down by grief, inconsolably alone, she went to Boston in search of a job. Soon she wrote her family a baleful confession: she had sunk into such despair, she had considered throwing herself into the cold brackish water of the Mill Dam. Drawn to the promise of oblivion, she had stared at her shimmering reflection for a long time. "But then it seemed so mean to turn & run away before the battle was over that I went home, set my teeth & vowed I'd *make* things work in spite of the world, the flesh & the devil."

"Love and Self-Love," her first story published in the *Atlantic,* was a product of this heroic resolution. A cathartic melodrama, it tells of a sixteen-year-old orphan who marries someone twice her age, a middle-aged man in love with another woman. (Alcott's models here were *Jane Eyre, Bleak House,* and numerous other Victorian page-turners.) Effie Home, the story's heroine, strives to create the familial abode promised by her surname. Failing that, she tries to drown herself. Only after this dramatic event does her distant husband awaken to the needs of his bride, and the story ends with husband and wife learning to value mutual respect more highly than the passionate love neither feels.

Critics have read Alcott's tale as a veiled account of Louisa's relationship with her father, who became less stern and distant after his daughter confessed her desire to end her life. But the story also registers larger concerns of the era—for instance, the proper way to treat orphans. Orphans populate American fiction throughout the nineteenth century and into the early twentieth. They hunt whales, sail down the Mississippi, suffocate in society's houses of mirth. The figure is linked to the migratory experiences of Americans, who severed ties with European culture and sought to construct new lives and new social identities upon the blank slate of a New World. Effie Home, like other fictional orphans, suffers from an abandonment that is as much existential as it is parental—a spiritual desertion Melville alludes to in *Moby-Dick* when he writes, "Our souls are like those orphans whose unwedded mothers die in bearing them." "Love and Self-Love" is not only a sentimental love story, then, but also a tale of profound alienation, a work that dramatizes the collapse of prior beliefs and the feeling of bereavement that results. In this context it is worth noting that Effie's older husband resembles not so much Bronson as his friend and neighbor, Ralph Waldo Emerson, who had allowed the young writer carte blanche to his library and prompted her to declare that he was "the man who has helped me most by his life, his books, his society. I can never tell all he has been to me." Despite Emerson's gifts, Louisa felt detached from the transcendentalist's extravagant claims on behalf of the spirit, which did nothing in the here and now to relieve the hardships of daily life. Louisa and her mother took in sewing, tutored, did piecework to earn money—while the philosophical Bronson sat in his garden and blandly contemplated the universe. Effie's estrangement from her husband can be read in part as a separation from the idealism of her father's generation.

Once he reached home, Bronson presented Louisa with the March issue of the *Atlantic,* which he hoped would "encourage and lead her to some appreciation of the fair destiny that awaits her if she will be true to her gifts as she has begun."

Louisa was touched. "Though in many people's eyes Father may seem improvident, selfish, and indolent," she had once admitted in her journal, "—though he often does in my own and I wish he were more like other men—yet I begin to see the purpose of his life and love him for the patient persistence with which he has done what he thought right through all opposition and reproach, for that is what few do I find." Now Bronson returned the compliment. The last words he put into his journal on this particular evening were addressed to his daughter: "I am pleased, and proud of thee."

The publication of "Love and Self-Love" kindled Louisa's creativity. Within days she fell into the vortex again, spending the cold nights huddled over a candlelit desk while the rest of the family slept. Within a week she completed another story, this one more sensational than the last, which she immediately sent to James Russell Lowell at the *Atlantic.* More audacious, more daring than "Love and Self-Love," the new story flirted with the great taboo of nineteenth-century America. It addressed the unspeakable subject that preoccupied antebellum Americans more than any other: sexual relationships between whites and blacks.

Toward the end of February, Louisa summoned her mother and father into the parlor to read this new work, which was entitled "M. L., An Abolition Tale." She had finished the story days earlier, but the week had been too filled with excitement to unveil it until now. For one thing, Franklin Sanborn had just returned to Concord. He carried himself even more erect than usual, swaggering as if he had been leading the fight against slavery instead of hiding in Canada for the past month. And he brought with him two guests he had picked up on his way through North Elba, New York, where John Brown's surviving family still lived. Brown's youngest daughters, Sarah and Anne, were aged sixteen and thirteen, and Sanborn had persuaded them to attend his academy. "They are bright girls," he wrote, "though of course unused to the ways of society." Henry David Thoreau's mother and sister hosted a welcoming party for the girls, and before long the Brown sisters were boarding with the Alcotts, where they played cribbage, chess, and a card game called casino Morris with Louisa and Abba in the evenings.

Years later Anne Brown recalled those times with bittersweet fondness. She still grieved for her father and three brothers, all of whom had been killed at Harpers Ferry or executed soon after. As it turned out, John Brown's trial had proved far more effective for the cause of abolition than his actions in Kansas or at Harpers Ferry. His defense lawyer attempted to enter a plea of insanity, but his client refused. Brown wished to testify on his own behalf during the trial. "This court acknowledges, as I suppose, the validity of the law of God," he announced on the day of his sentencing. "I see a book kissed here which I suppose to be the Bible, or at least the New Testament, that teaches me that all things whatsoever I would that men should do to me, I should do even so to them." His raid upon the federal arsenal had been an endeavor "to act upon that instruction. . . . I believe that to have interfered as I have done—in behalf of His despised poor, was not wrong, but right. Now, if it is deemed necessary that I should forfeit my life for the furtherance of the ends of justice, and mingle my blood further with the blood of my children and with the blood of millions in this slave country whose rights are disregarded by wicked, cruel, and unjust enactments—I submit; so let it be done!"

Anne Brown missed her father, but she was also glad to be inside a warm home in a town that was far removed from the grim subsistence of North Elba. Concord was lively and prosperous. In February there had been a masquerade party for the young people. Ellen Emerson dressed as an old woman, and her younger sister, with whom it was rumored that Sanborn was infatuated, dressed as a Dickens character. Someone else dressed as Topsy, the unruly slave girl in *Uncle Tom's Cabin,* and Louisa appeared as a nun, with a large black cross hanging from her neck. There was something prophetic about the costume, Anne thought, for later that spring, when a suitor began courting Louisa, she asked her new friend why she didn't marry him. Louisa replied, "Ah, he is too blue and too prudent for me, I should shock him constantly."

Louisa's new story was designed to shock the prudent. It was based on a real-life story she had heard from her uncle Samuel J. May about a biracial man and his white wife. William G. Allen was the child of a mulatto mother and a white father; raised by a free black family, he eventually made his way to Boston, where he clerked for Ellis Gray Loring (another relative of Charles Loring Brace) and lectured on abolition, the equality of the races, and the inevitability of "amalgamation." Soon he became a professor of Greek and German literature at a small college in upstate New York, where he fell in love with a white student named Mary E. King. The couple's engagement prompted outrage in the community; threats were made on Allen's life. As he described it in his pamphlet *The American Prejudice Against Color,* he narrowly escaped a mob "armed with tar, feathers, poles, and an empty barrel spiked with shingle nails."

The events described in Allen's pamphlet had been simmering in Louisa's imagination for months. They mingled with tales of unrequited love, with the raging debate over slavery, and with some of Louisa's favorite chapters in *Uncle Tom's Cabin.* Allen's autobiographical pamphlet seems to have put Louisa in mind of George Harris, the mulatto character in Stowe's novel who runs away from servitude and vows to die rather than return to slavery. George stands in sharp contrast to the titular Tom, humble and Christlike, whom Stowe meant to represent as a redeemer of the nation. In the logic of the novel, George's resistance is the inevitable product of his Anglo-Saxon blood. His white parentage links him to the Founding Fathers and to their willingness to fight tyranny and oppression. But if George has inherited the revolutionary ardor of his white forefathers, his eagerness to exact vengeance on his master is all too reminiscent of Toussaint L'Ouverture, Nat Turner, John Brown, and other leaders of slave rebellions. Disguised as a Spaniard and tracked by a slave catcher in the first half of the novel, George determines to kill his pursuer in the name of freedom. Stowe sidesteps this incendiary scene of interracial violence by having George merely wound his pursuer. But at the end of her book she banishes the rebellious character from her imagined America, "colonizing" George and his family in Africa.

Another influence on Louisa's story was surely *On the Origin of Species*. By this time she knew of the book from her father's recent conversations, as well as from Franklin Sanborn and especially Henry David Thoreau, who discussed it on repeated occasions throughout the first half of the year. Louisa probably had some idea of the ongoing quarrel between Asa Gray and Louis Agassiz, since the Swiss naturalist was a friend of Bronson's and the two men regularly dined at Saturday Club meetings. (Later that summer another one of Louisa's stories would share space in the *Atlantic* with Gray's essay on Darwinism.) But beyond all these connections she would have been interested in Darwin's work for the same reason the Concord transcendentalists were: because it provided a rigorous scientific argument that suggested that all people were linked by inheritance and destined for progressive improvement. These ideas are at the heart of Louisa's abolitionist story.

Paul Frere, or "Brother Paul," is the romantic foil for the heroine Claudia, whose name echoes the Gentile woman addressed in the biblical Paul's second letter to Timothy. Like her scriptural antecedent, she will hear a gospel of love and learn to believe. In Louisa's story, Claudia hears Paul singing a classical oratory in an adjacent room at a party. Even before she sets eyes upon him, she has fallen in love. Inquiring about the stranger, she learns that Paul is, like her, an orphan, as well as "a Spaniard, and of noble family." When at last she is introduced to him, she is struck by his appearance: "Black locks streaked an ample forehead, black brows arched finely over southern eyes." Something about his appearance reminds her of "a picture she had often pondered over when a child of a tropical island, beautiful with the bloom and verdure of the South." Palm trees and orange groves populated the picture, "but looking nearer, the eye saw that the palm's green crowns were rent, the vines hung torn as if by ruthless gusts, and the orange boughs were robbed of half their wealth. . . . Far on the horizon's edge, a thunderous cloud seemed rolling westward, and on the waves an ominous wreck swayed with the swaying of the treacherous seas."

The imagery here is conventional, a typical representation of the slave-holding South. It echoes, for instance, the lush backdrop of Longfellow's "The Quadroon Girl," a poem Alcott knew by heart, in which "odours of orange-flowers, and spice" reach the Slaver in "the broad lagoon." (Several lines later in the poem a slave-owning father sells his mixed-race daughter for gold.) But Louisa's language contains something Longfellow's poem doesn't: a hint of humid sexuality. And the portentous storm in the distance, blowing wrack toward a troubled western shore, is a rich amalgam of suggestions—of the illicit interracial love that produced Paul, of Paul and Claudia's incipient romance, of the global slave trade more generally, and finally of the building storm between North and South.

Smitten by Paul, Claudia never suspects his racial inheritance until a jealous friend reveals his secret, literally branded on the palm of his hand. (The initials M. L. stand for Maurice Lecroix, Paul's former master.) Paul grew up in Cuba, the illicit love-child of a planter father and a beautiful mixed-race mother. While his father lived, the boy was "lifted up into humanity," a cherished member of the household. After his father died, however, he was sold to a "hard master" and "cast back among the brutes." At this humiliating turn of events, Paul's European blood rebelled. "I could not change my nature though I were to be a slave forever," he confides to Claudia. Inherited from his father, like a feverish disease, are his "free instincts, aspirations, and desires."

Louisa had received some of her ideas about inheritance from her father. Obsessed with genealogy, Bronson spent decades tracing the ancestral Alcocke family to its Puritan roots. He believed physical and spiritual qualities were passed from generation to generation—that people inherited even their moods and temperaments from their forebears. He also believed one's complexion and hair color were outward symbols of inward qualities.

This meant that dark-haired and dark-complexioned people were passionate and moody; those with lighter complexions were intellectually and spiritually superior. Bronson, who had blue eyes and blond hair, liked to tell Louisa that she was a "true-blue May, or rather a brown," alluding to Abba's olive skin and brunette hair. By the early 1850s he had constructed a private taxonomy of race that ranked types of people in ascending order from dark to copper to yellow to white.

These ideas disturbed his friends, who saw them as racist. When the reformer Ednah Littlehale Cheney told Bronson that the Swedish scientist and mystic Emanuel Swedenborg considered blacks the most beloved of all races, he merely smiled and answered, "That is very nice of Mr. Swedenborg." Like many abolitionists of the era, Alcott had no trouble believing that black people deserved freedom while also assuming that God had made them inferior for obscure reasons of His own.

Louisa engaged her father's ideas about inheritance and race in "M. L., An Abolition Tale," but the story addressed an even more pressing public question: whether black or mixed-race people could be integrated into Northern society through love and marriage. Louisa's literary hero, Harriet Beecher Stowe, had thought not: she banished every biracial character in *Uncle Tom's Cabin* to Africa, neatly solving the problem of amalgamation and interracial society. Stowe was in the mainstream. In 1862, when Abraham Lincoln met with a "Committee of colored men" to discuss the future, he expressed his concern that both races living side by side were bound to have "general evil effects on the white race." Worried that an intermixture of peoples would erode American society and spark resentment among whites, he concluded, "It is better for us both, therefore, to be separated."

Louisa's story was more radical in its prognosis. Her white heroine ignores all social prohibitions against interracial relationships, heeding instead a "Diviner love." Soon after Claudia agrees to marry Paul, she is wracked with doubt. She considers "the worldly losses she might yet sustain" and briefly despairs at her future prospects. But in the end she determines to commit "social suicide." Forsaking "the emptiness of her old life," "the poverty of old beliefs," she marries Paul.

This was a bold move. Sensationalist "tragic mulatta" stories often highlighted the exotic sexuality of biracial characters, but they usually portrayed that sexuality as coarse, animalistic, sometimes even predatory. Louisa ignored these conventions, violating social and literary propriety in an imaginative act that exceeded what her culture deemed acceptable. She created a happy and enduring marriage between her white and mixed-race characters—a marriage that would have been illegal in the South and would remain so for another century. For Louisa, mixed-race marriage implied the future: the slow unraveling of racial prejudice, the redemption of America through interracial love. In language that echoes the long-standing rhetoric of Northern abolitionism and places both a Christian and a Darwinian emphasis on common descent, Paul received "a welcome to that brotherhood which makes the whole world kin."

Written around the time John Brown's daughters arrived in Concord, "M. L." is a counternarrative to the violent raid at Harpers Ferry, a story that dramatizes how selfless love might produce racial equality more quickly and effectively than warfare. But this selfless love comes with a cost, especially for Paul. The moment Claudia sacrifices social respectability for love, he becomes "the weaker now." "'I accept the bondage of the master who rules the world,'" he tells Claudia, referring to the newfound Christianity he adopts as he marries. "As he spoke, Paul looked a happier, more *contented* slave, than those fabulous captives the South boasts of, but finds it hard to show."

This abrupt—and disappointing—ending may be Louisa's effort to render her story fit for publication. Always a keen observer of the literary marketplace, she knew that stories celebrating marriage between people of two races were forbidden from print. But the conclusion also hints at Louisa's own uncertainties about a future interracial society, which threatened to radically transform the stable and largely homogenous New England culture in which she had spent her entire life.

As it turned out, even with the symbolic emasculation of Paul, the story proved much too risqué for the *Atlantic*. "Mr——won't have 'M. L.,' as it is antislavery, and the dear South must not be offended," Louisa soon confided to her journal. "Mr——" was James Russell Lowell, an avowed abolitionist who nevertheless believed Louisa's work was too incendiary for national publication. The story would eventually see print some three years later, during the height of the Civil War, in a pro-Union periodical called the *Commonwealth* that was edited by Franklin Sanborn.

But in the early months of 1860, no major magazine would print Louisa's "Abolition Tale." After the execution of old John Brown, Southerners and Northerners had come to view each other not just as different peoples but as mortal enemies. Tempers simmered on both sides. In the North, most people decried Brown's actions but admired his Puritanical belief. In the South, many read about the bells tolling across New England on the day Brown was hung and assumed the worst. The rising Republican Party, they asserted, was clearly "organized on the basis of making war" against its Southern neighbors. And indeed, in the first half of the year, the nation seemed to lurch ever closer to open conflict over the question of slavery. In such an environment, even the most progressive literary journal in the nation was not quite ready for an interracial love story—at least one that did not end in tragedy.

13

Tree of Life

Asa Gray was having problems of his own with the *Atlantic* that spring. "I mailed you a clean copy of my Review," he wrote Darwin in February, referring to his forthcoming essay in the *American Journal of Science*. "I have sent it to Agassiz. He is childishly apt to be offended at any opposition, but I have, as you see been very careful to avoid all personal offence."

Two weeks later Darwin replied, declaring that Gray's review was "by far the most able which has appeared, & you . . . have done the subject infinite service." Throughout the international scientific community, Gray's thoughtful essay would be well received. From England, Joseph Dalton Hooker wrote his friend to congratulate him on his cogent explication of Darwin's theory. "You have succeeded *au merveille*—both with the Nat. Hist., Metaphysic, & above all the Theological aspect of the whole question," Hooker exclaimed. Francis Boott, an American botanist also living in England, wrote to say the *Journal* review was "the best yet on the subject."

But now, as New England's snows melted and Gray began to work on his series of articles for the *Atlantic,* he encountered a difficulty. He had never written for a popular audience before, and for the first time in his career, he faced a challenge that his rival Agassiz had mastered with such élan: presenting complicated ideas to an untutored audience with concision and clarity. Gray wanted to convey to his readers the intellectual adventure contained within the *Origin of Species*. He wanted to suggest how the book seemed to bring the world to life, to make it pulse with meaning and significance. This was difficult in and of itself. But first he had to win over an audience that was understandably skeptical of Darwin's bold claims.

He began his first *Atlantic* essay by comparing evolutionary theory to a new pair of pants: "sure to have hard-fitting places" but equally certain to become comfortable over time. After all, the theory had scientific precedent. Investigations into the way species succeeded one another had been a topic of scientific concern for years now. And this idea had been leading up to the most important question of all: "the question of their origin." Astronomers, Gray wrote, now believed that the solar system originated "from a common revolving fluid mass," and physicists "speculate . . . steadily in the direction of the ultimate unity of matter." Science, in other words, had been moving in the direction of developmental theories for some time.

Still, it was an all-too-human reaction to "cling to long-accepted theory just as we cling to an old suit of clothes." Darwin's book would surely oppress some readers, offering ideas that portend "no good to old beliefs." The theories in the *Origin* were not unlike Galileo's once-blasphemous observation that the Earth revolved around the sun. While the *Atlantic*'s readers "might have helped to proscribe, or to burn" the Italian scientist during the Renaissance, they certainly would have recovered their "composure, [after they] had leisurely excogitated the matter," eventually coming around to Galileo's view. Some might even have granted that "the new doctrine was better than the old one, after all, at least for those who had nothing to unlearn."

Gray then cheerfully waded into the polygenist controversy, asserting that Darwin's book offered refreshing clarity "when we consider the endless disputes of naturalists and ethnologists over the human races, as to whether they belong to one species or more, and, if to more, whether three, or five, or fifty." He admitted that Darwin's vision contradicted transcendentalist philosophy and idealist science, both of which understood nature as a repository of divine values. In the *Origin,* "all Nature is at war, one species with another, and the nearer kindred the more internecine." He even touched on the book's unstated premise that humans were related to animals, admitting that "it is only the backward glance . . . that reveals anything alarming." If Darwin's hypothesis "makes the negro and the Hottentot our blood-relations—not that reason or Scripture objects to that"—then the "next suggests a closer association of our ancestors of the olden time with 'our poor relations' of the quadrumanous family than we like to acknowledge." Gray was referring to apes, of course, and he rushed to assure his audience that no evidence of a "missing link" between humans and primates existed in the fossil record. After all, one item in particular separated humans from every other creature on the planet, and that was language.

One of the most famous passages of the *Origin* begins as follows: "The affinities of all the beings of the same class have sometimes been represented by a great tree." This tree was abundant with "green and budding twigs" that "represent existing species." It also carried dead limbs and boughs that "represent the long succession of extinct species." Darwin was suggesting that if we could rewind the spool of time, traveling far enough back to a period before human existence, before the age of dinosaurs and the cycles of glacial subsidence and the upheaving of volcanic mountains, we would discover a single parent from which all living creatures had descended: a common root from which the Tree of Life had sprung. For "all living things have much in common: their chemical composition, their germinal vesicles, their cellular structure, and their laws of growth and reproduction."

The Tree of Life was composed of winners and losers: budding boughs that had adapted to their conditions, dead branches that had not. And if one applied this concept to humans—as nearly every early reader of Darwin's book did—it suggested not only the profound connection linking all people, it also helped explain why some individuals flourished while others did not, even why entire *cultures* flourished and others died out. Still, there was one fairly significant problem with the Tree of Life—it lacked confirmatory evidence. In 1859 there was no way for Darwin to connect the Tree's most recent buds to its ancient trunk or roots; genetic science did not yet exist, and there were no tests to determine lineage and affiliation. Moreover, the fossil record was far more incomplete at the time than it is today.

Darwin turned to human language to illustrate the problem. "If we possessed a perfect pedigree of mankind," he said, "a genealogical arrangement of the races of man would afford the best classification of the various languages now spoken throughout the world." If scientists examined "all extinct languages, and all intermediate and slowly changing dialects," they would almost certainly discover that while "some very ancient language had altered but little, and had given rise to few new languages," others, "owing to the spreading and subsequent isolation and states of civilization of the several races," would have "altered much and given rise to many new languages and dialects." If languages evolved in the same way Darwin believed plants and animals evolved, linguistics might reveal a common human ancestry connecting "together all languages, extinct and modern, by the closest affinities, and would give the filiation and origin of each tongue."

Gray borrowed this idea for his first essay in the *Atlantic*. In the future, he wrote, naturalists might determine relationships between species, "just as philologists infer former connection of races . . . [from] generic similarities among existing languages." Gray was making a specific point about scientific evidence and theories: that one could draw inferences from incomplete evidence so long as one was willing to revise those inferences when new evidence became available. But sometime that summer Charles Loring Brace opened his copy of the *Atlantic* to read Gray's review and came across the passage about languages. That passage suddenly crystallized all of the diverse data he had collected in his ethnological research, giving his book project a sense of direction and purpose it had previously lacked.

Earlier that year a visitor had appeared at the office of the Children's

Aid Society: a wizened black boy, solemn and shy. Brace sat the child down and learned that his name was Edward. That his parents were drunkards. That he had worked odd jobs in a soap factory, had carried water in a saloon, had spent the previous night huddled in a gashouse in Yonkers—that he was one more vagrant child in a city swarming with vagrant children. Then Brace looked more closely. Something prompted him to suggest the boy be given a bath. When Edward returned to his office, his face was no longer "disfigured with smoke and tar." His clothes were new, his face was white, and it turned out that he was in fact an Irish orphan. "The transformation was so effectual," Brace related, "that he was some time seated near one of his late companions, before he was known as the ex-colored boy."

Brace loved this story. He wrote about Edward several times in 1860, his narrative growing with each repetition. He was especially fond of the child's explanation of his condition: "Last night, I slept in the Gas House at Yonkers, and that was how I was metamorphosed into a colored boy"— as if being a black person were simply a dream, the product of an infernal nightmare, a transient state to be scrubbed away in the clear light of morning. Brace also liked the story because he wished it were true for society as a whole. He wished such simple transformations from black to white were possible, for if they were, they would provide a convenient solution to the most pressing problem of his age.

From a young age Brace was convinced that slavery was abhorrent to God, because it demeaned a significant portion of His creation. As an adolescent he attended antislavery lectures by Wendell Phillips, the Boston abolitionist who made his name in the 1830s with an address entitled "No Union with Slaveholders." Phillips had survived lynch mobs and threats of assassination because of his beliefs, but he refused to remain quiet on the topic of slavery. He argued that the South and the North were different tribes worshiping different idols. "The trial of fifty years only proves that it is impossible for free and slave States to unite on any terms," he declared, "without all becoming partners in the guilt and responsible for the sin of slavery."

Brace emphatically agreed. He argued with his college friends, who considered him an extremist. If "Slavery is a *sin per se*," then it made sense to strike down any and all authority that supported it—the Constitution, the judicial courts, even those misguided and complicit churches that defended it through a selective reading of scripture. But like so many Northerners, he found it impossible to imagine an American society in which blacks and whites lived side by side in peace and comity. Not that he couldn't hope. Long before he encountered Darwin's *Origin,* he was convinced that people were susceptible of radical change and transformation, that human history was progressing toward perfection, that one day a foundational connection among all peoples would be recognized. This is why he had been so deeply stirred when he first read Asa Gray's copy. Darwin's statement that "natural selection can act only through and for the good" reinforced his belief in a Divine Providence that oversaw a universe inevitably conducing toward good. Brace was similarly drawn to Darwin's assertion that Nature was "silently and insensibly working, whenever and wherever opportunity offers, at the *improvement* of each organic being in relation to its organic and inorganic conditions to life." Natural selection relied on chance and random variation for its operation, Brace understood, but its outcome was the very opposite of accident. Directional and purposeful, natural selection was simply the way God enforced His will and vision through nature. Perhaps this progressive force would one day solve America's racial problems.

Shortly after he read Gray's first review in the *Atlantic,* Brace began to reorganize his manual of ethnology so that it focused on human language. Wanting to confirm Darwin's idea that all human races stemmed from the same branch, he spent the autumn reading nearly every work available in the comparatively new field of linguistics, once again taking notes, then beginning to draft chapters. Eventually he sent Gray a sample of his work.

"You are very welcome to such casual criticism as I can offer," Gray replied from Cambridge after reading the manuscript-in-process. He then proceeded to question a number of Brace's premises, including his contention that "savage" people invariably rose to the level of "civilized" people.

"The general fact of a segregated people . . . becoming best adapted to the particular climate, etc., through Natural Selection is clear enough," Gray explained. Africans had adapted to the climate of Africa, just as indigenous Native Americans had adapted to the plains and deserts of the Americas. The fact of racial differences could perhaps be explained as "the best adapted . . . surviving in the long run, and the peculiarities transmitted by close breeding." But from a biological perspective, adaptive advantages trumped what Brace referred to as "civilization." Societies with established institutions did not necessarily survive; history was filled with fallen empires. Brace had been thinking of Europe, particularly the way its immigrants had spilled into North America and flourished like some invasive plant for the past two hundred years. But if biology was any guide, Gray continued, the "foreign though conquering race would be less prolific and less enduring than the native, etc., etc." Which led him to ask, "If you cut off all future immigration into North America, would the Indians resume possession of the country? or else our descendants become a copper-colored race?"

Brace did not have an immediate answer to these questions.

Gray's essays for the *Atlantic* elicited a steady stream of letters from readers. Some responded with unalloyed enthusiasm for the new idea, which seemed to unlock the mysteries of nature. Others reacted with anger toward a theory that proposed to unravel something else—their belief in a divine Being who had placed humans at the summit of creation. One *Atlantic* reader wrote Gray to ask him to provide a countertheory that would combine the old notion of special creation with the newer modification through descent: "There is danger lest with the tendency of modern science to simplicities of generalization, this very essential principle of the Divine work" would be lost. Couldn't it be true that "a single great harmony may include many apparent discords, many diverse modes of operation"?

Gray was flattered by the response, but he understood his immediate task as making Darwin's theory understandable to a general audience—not as offering alternatives to that theory. And he was running into a problem that sooner or later afflicted nearly every serious reader of the work. The tone of Darwin's book—so reserved, so reasonable—cloaked insights that were explosive and deeply unsettling. Gray had recognized these insights when he first read the book, but as with most radically new ideas, they had taken a while to register. Increasingly now he saw that Darwin's book resembled Copernicus's *De Revolutionibus* or Galileo's *Dialogue Concerning the Two Chief World Systems*. Like them, the *Origin* threatened to diminish humanity's place in the universe. It suddenly made obsolete a comforting way of looking at the world, transforming not only human history but the way people thought of themselves in relation to God. And even Gray wasn't entirely at ease with that realization.

14

A Jolt of Recognition

With the possible exception of Asa Gray, no American read the *Origin of Species* with as much care and insight as Henry David Thoreau. Throughout the first week of February, he copied extracts from the *Origin*. Those notes, which until recently had never been published, comprise six notebook pages in a nearly illegible scrawl. They tell the story of someone who must have read with hushed attention, someone attuned to every nuance and involution in the book. In their attention to detail, they suggest someone who assiduously followed the gradual unfolding of Darwin's ideas, the unspooling of his argument, as though the book of science were an adventure tale or a travel narrative.

He was drawn to Darwin's compendium of facts, which illustrated the delicate interplay of causes leading to the survival or extinction of species. Darwin wrote, "The number of humble-bees in any district depends in a great degree on the number of field-mice, which destroy their combs and nests." Thoreau copied the sentence into his notebook, probably because he enjoyed the cause-and-effect relationship it implied. He had always been interested in the quirky, arcane detail. "Winged seeds are never found in fruits which do not open," he read in the *Origin*, transcribing the sentence into his natural history book. He recorded the strange (if incorrect) statement that "cats with blue eyes are invariably deaf," something Darwin had gleaned from a work on zoological anomalies by Isidore Geoffroy St. Hilaire, who mistakenly assumed that *all* blue-eyed cats were deaf rather than the majority, as is actually the case.

He also admired Darwin's genius for experimentation. Thoreau had described his own efforts in *Walden* to disprove the local myth that the pond was of unusual depth. With a stone tied to the end of a cod line, he "could tell accurately when the stone left the bottom, by having to pull so much harder before the water got underneath to help me"—a procedure that enabled him to chart the pond's topography and discover its shallows and depths. He had even provided a map for interested readers. Now he discovered a similar impulse in Darwin. The British naturalist wanted to determine how far birds might transport seeds caught in their muddy feet; this would explain how identical plant species might be found thousands of miles apart. From the silty bottom of a pond near his home he procured some "three table-spoonfuls of mud," which "when dry weighed only 6¾ ounces." He kept the mud in his study for six months, "pulling up and counting each plant as it grew; the plants were of many kinds, and were altogether 537 in number; and yet the viscid mud was all contained in a breakfast cup!" The charm of the experiment resided in its simple ingenuity; from common household items Darwin had made a marvelous discovery: 537 plants!

Thoreau was most urgently drawn to Darwin's *ideas*. That the struggle among species was an engine of creation struck him with particular force. It undermined transcendentalist assumptions about the essential goodness of nature, but it also corroborated many of Thoreau's own observations. While living on Walden Pond, he had tried to discover the "unbroken harmony" of the environment, the "celestial dews" and "depth and purity" of the ponds. "Lying between the earth and heavens," he wrote, Walden "partakes of the color of both." But sometimes a darker reality intruded upon this picture. "From a hill-top you can see a fish leap in almost any part; for not a pickerel or shiner picks an insect from this smooth lake but it manifestly disturbs the equilibrium of the whole lake." Something portentous and uneasy lurks about this sentence. The "simple fact" that animals must consume other animals to survive upsets Thoreau; it disturbs the equilibrium of one who wishes to find harmony and beauty in his surroundings. Thoreau tries to laugh it off, calling the dimpled lake the result of "piscine murder." Yet Darwin provided an explanation for nature's murderous subtext. Competition and struggle influenced "the whole economy of nature." It drove species to change and adapt. It *created*. It was the cost of doing nature's business.

By the time he finished the first chapters of Darwin's book, Thoreau had seized upon two of its principal ideas. The first was variation: the essential building block of natural selection. Variation received more attention than any other topic in the *Origin,* mainly because it provided the "means of modification" that enabled organisms to adapt. In the past Thoreau had considered variation from a transcendentalist's perspective. "I expected a fauna more infinite and various," he noted with disappointment one spring day, "birds of more dazzling colors and more celestial song. How many springs shall I continue to see the common sucker (*catostomus Bostoniensis*) floating dead on our river! Will not Nature select her types from a new fount?" Written five years before Darwin's book was published, the passage reveals a basic tenet of transcendentalist philosophy: Thoreau expects Nature to answer to the demands of his imagination, to serve human needs. But the dead fish he finds each spring in the Musketaquid River also suggested to him an imperfect fit between species and environment. Calling for more "types from a new fount," Thoreau personifies Nature as a printer, able to produce new *fonts* with the prodigality of a poet who has sipped from the *fountains* of Hyperion. In this characteristic pun, Thoreau anticipated Darwin, suggesting that differences among individuals might eventually result in the diversification of species.

Following this line of thought now, he copied a number of passages about variation and hereditability, noting for instance the puzzling phenomenon of young horses born with stripes on their shoulders. Why did those stripes disappear as they aged? "How simply is the fact explained," wrote Darwin, "if we believe that these species have descended from a striped progenitor, in the same manner as the several domestic breeds of pigeon have descended from the blue and barred rock-pigeon!" (Darwin's point was that the history of a species was encoded in the body, that physical characteristics provided clues about relationships to ancient progenitors.) Thoreau also copied another remark under the heading "*Variability of flowers,*" which stated that species belonging to the same genus tended to share characteristics, such as variations in color, while those belonging to separate genera did not.

What really interested him, however, was Darwin's discussion of geographical distribution—the same topic that had engaged Asa Gray a decade earlier. In 1850 Thoreau had noticed a pine seedling in his yard, miles from any other pine, prompting him to wonder how it had gotten there. He began to study the way squirrels transported nuts and seeds from one location to another; then he followed the aerial voyages of milkweed spores and dandelion seeds. Soon he was observing cockleburs and other barbed seeds that attached themselves to animals and clothing, and for a while he considered whether the railroad might play a role in dispersing nonnative seeds to new locations. The question he was trying to answer was one he had asked in *Walden:* "Why do precisely these objects which we behold make a world?" Thoreau wanted to understand how the oak and pine woods surrounding Concord had sprung into existence. Why did this locale support robins and butternuts rather than, say, parakeets and pecan trees? How had each living thing come to inhabit its particular spot on the planet?

By concerning himself with this topic, he was at the forefront of natural science. Alfred Russel Wallace, whose wide-ranging travels had given him keener insight into the distribution of living things than almost anyone else in the world, would eventually explain the significance of the problem in his book *Island Life* (1880): "We can never arrive at any trustworthy conclusions as to how the present state of the organic world was brought about until we have ascertained with some accuracy the general laws of the distribution of living things over the earth's surface." In his own travels on the *Beagle,* Darwin had discovered that physical barriers—oceans, deserts, and mountains—often confined plants and animals to highly circumscribed regions. The Galápagos Islands were but one example. Sometimes an identical species was scattered across distant continents, even across vast oceans—as in the case of the Japanese flora found in eastern North America. While Louis Agassiz claimed that such examples implied separate and divine creation, a consensus was beginning to emerge within the scientific community that species were migratory and dynamic, settling wherever climate and resources facilitated their growth.

Darwin was convinced that the distribution of plants and animals shed light on evolutionary development, especially when a species became isolated and developed on its own. Some of the most delightful passages in the *Origin of Species* describe the experiments he conducted to determine how organisms scattered across the globe. He immersed seeds in saltwater for months at a time, then planted them to see if they would grow. He calculated the distance these seeds might travel across the ocean while immersed. (He determined that some could travel as much as 924 miles by prevailing Atlantic currents.) As for animals, he placed duck feet in a tank of water containing minuscule freshwater snails to see if the tiny creatures would take hold of the webbing. Revisiting his old travel notes, he discovered that a water beetle blown onto the deck of the *Beagle* had traveled some forty-five miles; this led him to speculate how insects and birds might cover enormous distances during a gale.

These experiments not only revealed how the world might have become populated; they also suggested just how *accidental* was that process. Far from the carefully organized scheme Agassiz and other special creationists described, Darwin's world was the product of random and haphazard occurrences. The seeds of plants blew wherever the wind took them and germinated wherever there was enough sun and moisture. Animals followed land bridges or were swept away by flash floods or hurricanes; they were isolated on tiny islands in the middle of the ocean. Nothing was predetermined, nothing organized by design. General laws might govern these actions, but at the individual level, chance prevailed.

Because he was already interested in the topic, Thoreau transcribed more passages from Darwin's chapter on geographical distribution than from any other in the *Origin*. He carefully noted Darwin's assertion that "I have not found a single instance, free from doubt, of a terrestrial mammal (excluding domesticated animals kept by the natives) inhabiting an island situated 300 miles from a continent or great continental island—." And he meticulously followed Darwin's argument that isolated islands might produce special evolutionary conditions. "[The French naturalist] Bory St. Vincent long ago remarked that Batrachians (frogs, toads, newts) have never been found on any of the many islands with which the great oceans are studded," Darwin wrote. "I have taken pains to verify this assertion, and I have found it strictly true. I have, however, been assured that a frog exists on the mountains of the great island of New Zealand."

Thoreau wrote down this passage and appended a remark that showed just how thoroughly he had absorbed the intricacies of Darwin's discussion: The frog, he asserted, was surely "spawned not there."

Ultimately it was Darwin's *method* that left the deepest impression on Thoreau. The book was infused with a point of view: humorous and humane, stubbornly rigorous, breathtaking in its originality. As Thoreau pored over the *Origin*, he encountered many of his own thoughts packaged and reformulated in a style that was not just scientific but something we would now call *Darwinian*. In *Walden*, Thoreau had described a natural world prodigious in death: "I love to see that Nature is so rife with life that myriads can be afforded to be sacrificed and suffered to prey on one another; that tender organizations can be so serenely squashed out of existence like pulp,—tadpoles, which herons gobble up, and tortoises and toads run over in the road; and that sometime it has rained flesh and blood!" Thoreau was again trying to place nature's profligate waste within a larger philosophical context. Describing the stench of a dead horse "in the hollow by the path to my house," he claimed that the fetid atmosphere of decomposition indicated "the strong appetite and inviolable health of Nature."

Darwin's theory was grounded on similar observations. Nature might be responsible for countless "exquisite adaptations" and "beautiful diversity," as he put it, but beneath those adaptations and diversity was an incessant struggle. "Every one has heard that when an American forest is cut down," he wrote, "a very different vegetation springs up. . . . What a struggle between the several kinds of trees must have gone on during long centuries, each annually scattering its seeds by the thousand; what war between insect and insect—between insects, snails, and other animals with birds and beasts of prey—all striving to increase, and all feeding on each other or on the trees or their seeds and seedlings, or on the plants which first clothed the ground and thus checked the growth of the trees."

This picture of strife and competition is similar to Thoreau's version of a natural world, "so rife with life that myriads can be afforded to be sacrificed and suffered to prey on one another." But while Thoreau thought nature's monumental destruction was necessary for its health, Darwin was less certain. "Throw up a handful of feathers, and all must fall to the ground according to definite laws," Darwin exclaimed, "but how simple is this problem compared to the action and reaction of innumerable plants and animals which have determined, in the course of centuries, the proportional numbers and kinds of trees now growing on . . . old Indian ruins!" Nature was neither fable nor allegory; it was, rather, a handful of feathers falling as randomly as gravity allowed.

Years earlier, when he was still living near the pond, Thoreau had been walking along the steep bank of the Fitchburg Railroad on the first warm day in spring when he noticed that the south-facing side of the embankment was thawing in the sun. "Innumerable little streams" of the softened sand and clay "overlap and interlace one another," he observed, "exhibiting a sort of hybrid product, which obeys half way the law of currents, and half way that of vegetation." The more he looked, the more interested he became in this odd sand foliage: "As it flows it takes the form of sappy leaves or vines, making heaps of pulpy sprays a foot or more in depth, and resembling, as you look down on them, the laciniated lobes and imbricated thalluses of some lichens." Looking more closely, he was "reminded of coral, of leopards' paws or birds' feet, or brains and lungs or bowels, and excrements of all kinds."

Generations of English students have puzzled over Thoreau's unbridled enthusiasm for melting mud. His mounting excitement stems from a sense that he has cracked Nature's code, has "stood in the laboratory of the Artist who made this world and me" and caught a glimpse of the source of all creation. This way of looking at nature is of course deeply romantic. Thoreau was influenced by Goethe's *Italian Journey,* in which the German had theorized about a primordial "ur-plant" that served as the model of all subsequent plants. Goethe in turn had been influenced by Platonic forms; like Louis Agassiz, he believed there was an original, ideal plant upon which all subsequent plants were fashioned: "Otherwise, how could I recognize this or that form *was* a plant if all were not built on the same basic model." Thoreau extended the analogy. For him the sand foliage represented "the original forms of vegetation"; it revealed nature's blueprint. Then he made a leap: perhaps the flowing, branching mud revealed "the bony system, and in the still finer soil and organic matter the fleshy fibre or cellular tissue" of animals. After all, "What is man but a mass of thawing clay?"

> You here see perchance how blood vessels were formed. If you look closely you observe that first there pushes forward from the thawing mass a stream of softened sand with a drop-like point, like the ball of the finger, feeling its way slowly and blindly downward. . . . Is not the hand a spreading *palm* leaf with its lobes and veins? The ear may be regarded, fancifully, as a lichen. . . . The nose is a manifest congealed drop or stalactite. The chin is a still larger drop, the confluent dripping of the face.

To his credit, Thoreau maintained a hint of skepticism throughout this rhapsodic passage. Riffing on words that sound like the mucky slithering of sand, he was really poking fun at himself, as if to say: *Woe to anyone who embraces a theory for the tidiness of its explanation.* But there is also genuine excitement at the prospect of discovering a law that might explain organic life.

Which is why Darwin's conclusion surely made such an impression on him. Imagine the jolt of recognition Thoreau must have experienced when he came upon the book's final paragraph:

> It is interesting to contemplate an entangled bank, clothed with many plants of many kinds, with birds singing on the bushes, with various insects flitting about, and with worms crawling through the damp earth, and to reflect that these elaborately constructed forms, so different from each other, and dependent on each other in so complex a manner, have all been produced by laws acting around us. . . . Thus, from the war of nature, from famine and death, the most exalted object which we are capable of conceiving, namely, the production of the higher animals, directly follows. There is grandeur in this view of life, with its several powers, having been originally breathed into a few forms or into one; and that, whilst this planet has gone cycling on according to the fixed law of gravity, from so simple a beginning endless forms most beautiful and most wonderful have been, and are being, evolved.

This is Darwin the visionary, rejoicing in nature's profusion, its lush fecundity. The passage celebrates the complex and interdependent relationships among species that have developed over time, and its language vibrates with excitement—the same tremulous wonder with which Darwin recounted his youthful discovery of a new variety of beetle.

We tend to think of Darwin's theory as one of grim determinism, of pointless change and purposeless death. But to do so is to miss a crucial point about his thinking during the period in which he wrote the *Origin*. In 1859 Darwin not only admired the natural world's plenitude and capacity for transformation. He also believed that life's messy process, its extravagant creation and destruction, led to something worth celebrating: "the production of the higher animals"—including, of course, humans.

Darwin's portrait of a teeming, pulsating natural world deeply resonated with Thoreau. The *Origin of Species* revealed nature as process, as continual *becoming*. It directed one's attention away from fixed concepts and hierarchies, toward movement instead. It valued moments of evanescent change above all others. If it endowed each organism with a history, it also pointed to a future that was impossible to predict.

For Thoreau, this aspect of the *Origin* seemed to finish a sentence he had long been struggling to articulate. Once uttered, that sentence seemed to snap the natural world into place. Reading the *Origin,* Thoreau discovered someone else who understood nature as he did: abounding and vibrant, each niche swarming, each interstice filled with life, each living thing a small part of constant change, a participant in struggle and development, brimming with potential and significance.

Throughout the late winter and early spring of 1860, he continued his daily walks, his diligent measuring and collecting. He spent the cold New England evenings hunched over his journal. But now his prose kindled with a new energy. And something else happened. For eight years Thoreau had patiently compiled mountains of phenological data—information about the timing of nature's seasonal events. Over the years he had noted the flowering dates of hundreds of plants and recorded what environmental scientists now refer to as "leaf-out" and "ice-out" dates: that brief period when trees first show their leaves and when the ice on ponds melts.

Immediately after finishing Darwin's book, Thoreau began the tedious task of extracting and collecting this information. He reread his journals—thousands of pages—and copied the relevant information onto random slips of paper. A receipt from the family pencil-making business, for instance, became the repository of phenological information from 1852. Snow levels from 1854, which Thoreau recorded with a notched walking stick, were written on another scrap. As he carefully combed through his journals, he placed an *X* by each entry he transcribed.

When he was finished with this monumental task, he found he had copied information about more than one hundred trees and some sixty shrubs. He had described the height of grasses, the size of red maple leaves in May, the dates during which the "leaves of goldenrod [were] obvious." He recorded the growth of fir trees, of larches, the leafing-out of the fever bush, waxwork, red cedar, tupelo, red currant, poison sumac. He noted the day in which "chicadees have winter ways." He entered the date on which he first noticed the scent of decay.

With this process complete, he gathered his slips of paper and transcribed the information once again. This time his data went into a series of spreadsheets. (The term *spreadsheet* was not in use then; it is possible Thoreau invented a prototype.) These sheets of paper were nearly the size of a newspaper. On each one Thoreau listed a month, with a column devoted to every year from 1852 to 1860. Into these columns, in tiny, nearly illegible handwriting, he recorded all the information he had gathered.

What was he up to? The simple answer is that we don't entirely know. He may have been doing what scientists invariably do when their aggregated data become too large and unwieldy: organizing them into sets. But the painstaking work he began in 1860 enabled Thoreau to capture and quantify the processes of growth and death in nature—to discover patterns in nature's chaotic creativity. It also allowed him to determine if Darwin's theories held true in the natural environs he knew so well. In *Walden* he had precipitously leaped into wishful hypothesizing, drawing from a bank of thawing mud all sorts of conclusions about the human and the divine. The spreadsheets presented something radically different: a natural world sharply restricted to facts. If Thoreau hoped to find some law or principle that might unify nature, he now believed he first had to build a solid foundation of evidence.

Around this same time, he began incorporating ideas he derived from reading Darwin into a new lecture he was writing for the Concord Lyceum. That work was entitled "Wild Apples," and it is arguably the first piece of literature on either side of the Atlantic to be inspired by the theory of natural selection.

15

Wildfires

Bronson Alcott was disappointed to learn of Emerson's high opinion of *On the Origin of Species*. That spring at the Saturday Club, Louis Agassiz broached the topic of Darwin's book with Emerson, who apparently had already read portions of the *Origin* sometime after returning from his winter lecture tour. Most likely he obtained a copy at Thoreau's urging, dipping into the book, as was his custom, like a bee collecting pollen, flipping pages back and forth, seeking inspiration for his own essays. He almost never read a book from start to finish, but he had skimmed enough of the *Origin* to see that its premises could be absorbed into his own progressive notion of history and human aspiration and that it confirmed some of his own deeply held convictions about the universe.

Indeed, while Emerson would always think of Nature as a reflection of God's thought, he shared with Darwin a sense that the material world was best understood as fluid and ever changing. "There are no fixtures in nature," Emerson wrote in his 1841 essay "Circles." "The universe is fluid and volatile. Permanence is but a word of degrees." There was an endpoint to this motion, however: the creation of divinely inspired humans. "The continual effort to raise himself above himself," he continued in the same essay, "to work a pitch above his last height, betrays itself in a man's relations." One of Emerson's young followers, Moncure Daniel Conway, went so far as to say, "We who studied [Emerson] were building our faith on evolution before Darwin came to prove our foundations strictly scientific." Conway exaggerated, but not by much, when he noted that it had long been clear to Emerson that "the method of nature is evolution, and it organized the basis of his every statement."

Agassiz, on the other hand, had continued his very public campaign to discredit evolutionary theory. On February 15 he attended another meeting of the Boston Society of Natural History, where he listened impatiently as members reported on new bird species and read correspondence from far-flung collectors. Then he stood up and expressed his profound distaste for the new book. Darwin was a "successful writer and natural historian," he allowed, but his new theory, while admittedly "ingenious," was also "fanciful." *Fanciful* was one of Agassiz's favorite epithets: a quality insufficiently respectful of the careful thinking demanded by science. The possibility that species migrated from place to place instead of remaining in their allocated zone, in Agassiz's opinion, was *fanciful*. The notion that plants or animals changed over time was *fanciful*. The belief that whites and blacks shared the same ancestor was also most deplorably *fanciful*. Agassiz told the assembled society that Darwin actually seemed to have convinced himself that creation began "with a primary cell." From this improbable beginning, organic life had developed "by a process of differentiation and gradual improvement."

The great Swiss scientist then launched into a rambling exposition. He expressed his belief that animals presently inhabiting the Earth were wholly unrelated to similar species from the past. Previous creatures had been wiped out during cataclysms such as the Ice Age and had then been recreated by God. Moreover, no credible evidence existed to prove that animals had grown more complex over time, as Darwin insisted. "Animal representatives were as numerous and diversified in early geological periods as now," Agassiz said.

In subsequent meetings of the society, the geologist William Barton Rogers would skillfully dismantle most of these arguments. Rogers was the soon-to-be-president of the Massachusetts Institute of Technology. In his debates with Agassiz he employed an encyclopedic knowledge of North American geology to show that most of his opponent's claims were either false or partially true at best. But Agassiz cared little for debating the topic; he was not interested in debating a subject about which he was already certain. As far as he was concerned, his own theory of special creation rendered Darwin's null and void. This was one of the reasons he preferred discussing the topic with members of the Saturday Club, who were far more sympathetic to his ideas and who were also more likely to shape public opinion about Darwinian theory than a handful of scientific specialists.

During a spring meeting of the club, Agassiz waited until the dinner plates were removed and then, according to a visitor, "made some little fling at the new theory." Conversation around the table died off.

Emerson, who was sitting next to Agassiz, smiled. He looked directly at his friend and admitted that while reading Darwin's book, "he had at once expressed satisfaction and confirmation of what [Agassiz] has long been telling us." For the *Origin of Species,* in his opinion, seemed to authenticate all of "those beautiful harmonies of form with form throughout nature which [Agassiz] had so finely divined."

Agassiz was pleasantly surprised by this compliment. "Yes," he said, warming to the topic. He said something about nature's "ideal relationships," which were nothing less than the "connected thoughts of a Being acting with an intellectual purpose."

But Emerson didn't mean that exactly. At age fifty-seven, he had become too great a respecter of the material world, its obdurate presence in human affairs. For some years now his thinking had been consumed with the poignant limitations of experience. While he continued to believe that "the visible universe was all a manifestation of things ideal," he nevertheless acknowledged that the "visible universe" impinged with some frequency upon his own day-to-day life and in ways that were difficult to explain away. Might one not consider Darwin's new theory "a counterpart of the ideal development"?

Agassiz shook his head. "There I cannot agree with you," he replied, and quickly changed the subject.

Bronson Alcott attended this meeting of the Saturday Club, and although he remained mute on the topic in his journal, he instinctively sided with Agassiz. Interested in new ideas, Alcott was not deaf to the melody of Darwin's thinking; he just didn't care for its particular tune. The theory banished Mind from the universe, evacuated meaning from the cosmos. All his life Alcott had trusted that a wise and benevolent Soul animated creation, guiding it toward perfection. But about Darwin's theories he would later complain, "Any faith declaring a divorce from the supernatural, and seeking to prop itself upon *Nature* alone falls short of satisfying the deepest needs of humanity."

At the next meeting of the Saturday Club, he brought up the topic himself. Henry Wadsworth Longfellow and Oliver Wendell Holmes were in attendance, and the two poets sat flushed and buoyant as they enjoyed their wine and traded bons mots. At one end of the long table Bronson Alcott and Louis Agassiz sat huddled together, wrapped in conversation. The two men could not have been more dissimilar: Agassiz burnished with good food and wine, his manners expansively European, Alcott committed to his vegetables and water, his attention wavering now and then as he communed with some ineffable spirit.

Leaning forward, he told Agassiz that he was distressed by the theory of natural selection—by the way it *pretended* to explain how humans could have originated from lesser creatures. Alcott had been mulling over this aspect of Darwin's hypothesis. Emerson, in his famous essay "The American Scholar," had described humans as existing in a "grub state." Too many Americans were sleepy drones refusing to awaken, unaware of their innate capacity to transform themselves into spiritual butterflies. From Alcott's vantage, Darwin emphasized the animal nature—the *grub state*—of creation far too much. The English naturalist seemed morbidly attached to an amoral struggle of existence, which robbed humans of free will and ignored the promptings of the soul. Alcott had developed his own hypothesis about transmutation, which he wanted to share with Agassiz.

Franklin Benjamin Sanborn:
at the center of controversy.

Portrait of John Brown taken by the
African American photographer
Augustus Washington in 1846.

Charles Robert Darwin
around the time *On the Origin
of Species* was published.

ON

THE ORIGIN OF SPECIES

BY MEANS OF NATURAL SELECTION,

OR THE

PRESERVATION OF FAVOURED RACES IN THE STRUGGLE
FOR LIFE.

By CHARLES DARWIN, M.A.,

FELLOW OF THE ROYAL, GEOLOGICAL, LINNÆAN, ETC., SOCIETIES;
AUTHOR OF 'JOURNAL OF RESEARCHES DURING H. M. S. BEAGLE'S VOYAGE
ROUND THE WORLD.'

LONDON:
JOHN MURRAY, ALBEMARLE STREET.
1859.

The book that changed America.

Asa Gray one year before
arriving at Harvard.

Gray's botanical notebook.

Charles Loring Brace.

A magazine sketch of Five Points, one of the most impoverished and violent neighborhoods in America.

Ralph Waldo Emerson.

Emerson's "transparent eyeball," drawn by Christopher Cranch.

The most radical idealist in America: Bronson Alcott.

Thoreau attired for his daily saunter.

Louis Agassiz.

William Henry Johnson.

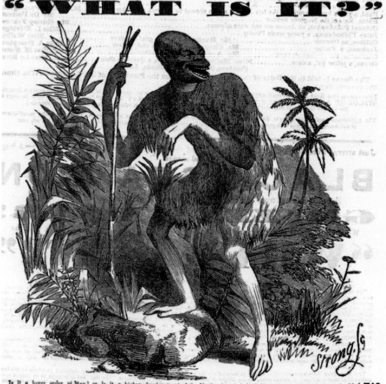

1860 newspaper advertisement for the What-Is-It?

Louisa May Alcott, at the desk her father built for her.

Arrest and rescue of Frank Sanborn.

A cartoon of Charles Darwin as an ape.

Thoreau's last photograph.

One of many cartoons portraying Lincoln as a primate.

"I have long desired," he said, "to bring my views of creation to the severest scientific test. To me the idea that man is the development from lower orders of beings is a subversion of the truth."

Agassiz grew attentive. According to a visitor that evening, he looked about the table "with a somewhat pleased glance at the rest of the company, whom he knew to be inclined to the hypothesis of Darwin."

"Yes, sir," Alcott continued, "an exact subversion of the truth. Man, I take it, was the first being; was he not?"

Agassiz looked confused.

Was it not obvious, Alcott continued, that "God could never have created a miserable, poisonous snake, and filthy vermin, and malignant tigers"?

Still puzzled, Agassiz asked who else might have created them.

"Must we not conclude," Alcott replied, "that these evil beasts which fill the world are the various forms of human sins? That when man was created they did not exist, but were originated by his lusts and animalisms?"

Alcott had inverted Darwin's evolutionary ideas. God had begun not with some lowly single-celled organism, he suggested, but instead with the *highest* form in nature—with humans. From there He had worked His way downward. Alcott told Agassiz he believed that *all* creatures had begun as humans, as part of a Universal Spirit, with some descending further into nature than others. The lower the animal in the chain of being, the further that particular animal had fallen from its true spiritual state.

At this point Agassiz thought it helpful to interject with some science. "But geology shows that these beasts existed many ages before man," he said.

"But may man not have created these things before he appeared in his present form?" Alcott earnestly asked.

Many years later Franklin Sanborn would speak of "that time when Alcott's ideas had become worn out." He was speaking of the years after the Civil War, when philosophical idealism was no longer in fashion and seemed increasingly a quaint relic from the past. In truth, that process had begun much earlier. Alcott's daughter Louisa revered his commitment to ideals, but she was also too honest to ignore her father's harebrained theorizing. She viewed her father with a combination of affection, resentment, and ridicule, much like Nathaniel Hawthorne, who moved next door that summer and satirized Bronson's obsession with diet. "He is now convinced," Hawthorne wrote, ". . . that pears exercise a more direct and ennobling influence on us than any other vegetable or fruit."

In 1860 Alcott's ideas were already becoming obsolete. Darwin's theory didn't make them that way, but it did hasten the process. The scientist's solid, unflappable language, his voluminous use of example and illustration—all these things exposed Alcott's abstractions as guesswork and conjectures, the unverifiable suppositions of someone who *needed* to believe in them.

Alcott's response to the theory of natural selection was to reject its materialism out of hand. At the same time, he borrowed its outlines so as to imagine a world filled with creatures that had descended from original perfection. In essence he applied Platonic ideals to evolutionary theory. Even Agassiz, the most idealistic scientist in America, understood that this approach was nonsense. He listened as long as he was able, and then he did what many people do when cornered by a monologist sharing a pet idea. He smiled politely and glanced at his watch.

Alcott failed to recognize the Darwinian references sprinkled throughout Thoreau's latest lecture. He thought "Wild Apples" "a celebration of the principles of Nature, exemplified with much learning and original observation: beginning with the Apple in Eden and down to the wildings in Concord." Alcott sat in the Town Hall and "listened with uninterrupted interest and delight" while Thoreau punned on Adam and Eve, on crab apples, on Johnny Appleseed. The lecture was a perfect example of Thoreau's ability to spin literary gold from the simplest materials.

By focusing on *wild* apples, Thoreau was making his standard argument for the uncultivated and untamable aspects of life. He traced the history of apple cultivation, sprinkling his talk with a decade's worth of facts and observations he culled from his notebooks and journals. Concord residents must have delighted to hear the village eccentric expound on the apple's place in Greek mythology and Homeric epic, to learn of the numerous binomial Latin names for apple varieties used by science, to hear pungent descriptions of the taste of different apples, including the acrid crabapple. They must have recognized with delight the "old farmer" Thoreau quoted as saying that apples in November "'have a kind of bow-arrow tang.'"

What they most likely did *not* notice was the influence of Darwin, which courses like a subterranean stream through the loamy prose of Thoreau's lecture. He traced the geographical distribution of apple trees "throughout Western Asia, China, & Japan." He described how animals helped disperse apple seeds, and he portrayed the fruit tree as an example of artificial selection, having been transmuted from an indigenous shrub to "the most civilized of all trees" by careful breeding over many generations. Referring to Darwin's discussion of dog breeders, he asked, "Who knows but like the dog, [the apple] will at length be no longer traceable to its wild original (No tree is more perfectly domesticated). It migrates."

By the time he delivered his lecture on "Wild Apples" in late February, Thoreau had long since finished reading Darwin's groundbreaking book. He continued to dwell on it, however, focusing especially on the book's third chapter, "The Struggle for Existence." Darwin's portrait of the "war between insects, snails, and other animals with birds and beasts of prey" captured a dynamic he had observed on his countless walks into the woods. But he was becoming more interested in the way this war also *linked* creatures together—something Darwin described as the way "plants and animals most remote in the scale of nature, are bound together by a web of complex relations." In Darwin's vision of nature, species and individuals honed themselves in strife. They came into being through continual friction with one another. "Many cases are on record showing how complex and unexpected are the checks and relations between organic beings," Darwin wrote, "which have to struggle together in the same country." Thoreau didn't express it in quite the same way, but he seems to have begun envisioning a natural world that resembled a democracy more than a kingdom, its citizens connected and yet perennially jostling for advantage. As winter came to a close, this fascination increasingly expressed itself in Thoreau's research into trees.

One of Darwin's examples stood out in particular—a passage in the *Origin* describing a Staffordshire estate. The land, which probably belonged to Darwin's father, was a large, barren heath. A generation earlier several hundred acres had been fenced off and planted with Scotch fir, the only pine variety native to Europe. Twenty years later the difference between the two areas was astonishing—"more than is generally seen in passing from one quite different soil to another," Darwin wrote. Twelve plants that did not exist on the heath flourished among the pines. Six insectivorous bird species, also wholly absent on the heath, lived there—implying a significant alteration in the insect population, as well. "Here we see how potent has been the effect of the introduction of a single tree, nothing whatever else having been done, with the exception that the land had been enclosed, so that cattle could not enter."

It was a simple but brilliant insight. By amassing details about plants and animals, Darwin had grasped how new environments might come into being. He could not trace every step that had created this new ecosystem—he had not observed it for the twenty years it took to develop—but he had a plausible theory to explain *why* the transformation had occurred. By introducing a new species to an established environment, humans had completely thrown off-kilter the dynamics of competition and coexistence, creating opportunities and disasters in its wake. An alteration in nature's equilibrium had introduced a chain of advantages and disadvantages to countless species, radically transforming the landscape.

Thoreau latched onto this particular moment in the *Origin* for several reasons. For one, it implied that the history of an environment was recoverable. If one accepted the premise that perpetual struggle between species led to the creation of *place,* then one could uncover its history and thereby determine why "precisely these objects which we behold make a world," as he had written in *Walden.* The passage in the *Origin* also reinforced the idea that such histories were provisional and unpredictable. Each living thing contained the potential for countless actions and reactions, and these in turn contained innumerable paths of development that were impossible to manage. Nature was *alive,* in other words, not static. And one other aspect of Darwin's story about the Staffordshire estate intrigued Thoreau: its *human* element. A completely new landscape had sprung into existence when a sentient being decided to introduce pine trees. This simple act had helped create a complicated environment, a new fact in the world. Thoreau had long suspected that people were an intrinsic part of nature—neither separate nor entirely alienated from it. Darwin enabled him to see how people and the environment worked together to fashion the world. Put another way, the *Origin* provided a scientific foundation for Thoreau's belief that humans and nature were part of the same continuum.

As winter waned, Thoreau walked with new purpose, looking for evidence of struggle and development in the woodlots just beyond the village. He spent time examining the margins between pine and oak forests, suddenly aware that these were the battle lines between two species that competed for the same soil and the same sunlight. He was more alert to the consequences of human activity. One day toward the end of March he left his mother's yellow house on Main Street and walked several miles to the nearby town of Acton, where a forest fire had recently consumed a thousand acres of timber. Here was an opportunity to observe a landscape wiped clean, an area that would have to start over from scratch—a laboratory of creation.

The air still smelled of smoke. The horizon was blurred with a thick, bluish haze. Thoreau stood before the smoldering waste, rapt and yet uncomfortable. Sixteen years earlier, in April 1844, in one of the most shameful acts of his life, he had accidentally burned down a large swath of the Concord woods. His campfire, built too large, had quickly spread beyond its pit. Thoreau jumped up to extinguish the blaze, first using his hands and feet, then grabbing a board from his canoe, "but in a few minutes it was beyond . . . reach; being on the side of a hill, it spread rapidly upward through the long, dry, wiry grass interspersed with bushes."

For years he could not bring himself to write about the event or the helplessness that had swallowed him as thick hot smoke spread through the woods and blackened the sky. There was something dreamlike and unreal about the experience. He watched as his beloved woods were consumed in a roaring red glow he had created. It was a lesson in hubris, in unintended consequences—a lesson in the way incendiary events engulf us and spread beyond our control. Thoreau could describe the incident only in a numb and distant voice. "The earth was uncommonly dry," he wrote, and the "fire, kindled far from the woods in a sunny recess in the hillside on the east of the pond, suddenly caught the dry grass of the previous year which grew about the stump on which it was kindled." He wrote nothing about the danger he had found himself in as the flames devoured every combustible material around him. Nor did he mention the strange, exciting beauty of the conflagration.

Now, sixteen years later, he stood outside Acton, before a blackened world. Charred trees lay amid embers, and the ashen hillside was silent and lifeless. It was cold—unseasonably cold, with the temperature reading thirty-one degrees, according to Thoreau's notes—and he saw traces of snow on gooseberry and lilac shrubs that were just beginning to bud some forty feet from the forest fire. It was "the dangerous time," he wrote that evening, a period of the year "between the drying of the earth, or say when the dust begins to fly, and the general leafing of the trees, when it is shaded again."

That night he wrote nothing about the world beyond Concord's woods—a world that seemed increasingly on the edge of combustion, too. A few months earlier he had been utterly consumed by the John Brown affair, unable to sleep, angry, and aflame with indignation. During the winter he had willed himself to quit thinking about the nation's baleful state of affairs, its compromises and hypocrisies, its unacceptable complicity with slavery. With little faith in the political process, which he believed favored the wealthy and self-interested, Thoreau read about the upcoming Republican and Democratic conventions with derision.

Darwin's book had helped him take his mind off these things. The *Origin* had redirected his thinking. It shifted his focus from a corrupt society that seemed incapable of reform to a natural world defined entirely by change and exuberant dynamism. In his journals and conversations, Thoreau still sometimes erupted in anger at Brown's unjust execution. But Darwin's theory soothed him during this period.

Which isn't to say he adopted the book unequivocally. That spring he continued to grapple with its unrelenting empiricism and with the inductive method of science more broadly. "Science in many departments of natural history does not pretend to go beyond the shell," he observed a few days before visiting the forest fire in Acton, "*i.e.*, it does not get to animated nature at all." For Thoreau, merely measuring and describing nature failed to capture its essence. Take the dog, for example. What was most interesting about the animal was "his attachment to his master, his intelligence, courage, and the like, and not his anatomical structure or even many habits which affect us less." Other aspects of the dog—its relationship to its kind, its fondness for warmth and touch, its interactions with people—conveyed core attributes far better than physical descriptions. Science missed the bigger picture. It failed to grasp what the ancient Romans would have called the *animus* of nature: its spirit, its mind, its purpose.

At times, Thoreau's thought bordered on the nostalgic. He longed for the transcendentalist's confidence in a natural world infused with spirit. He considered his increasing scientism an unwelcome sign of aging, as if the sap and vigor of youth were slowly petrifying. But he continued collecting data, continued filling his journal with notations on the arrival of the robin and bobolink, the budding of the spiraea and the Missouri currant. The third or fourth time he ventured to the wildfire site near Acton, he took notes on the moisture of the ground near the burned-out region. Then he returned to Concord.

On his way home he stopped to speak with a local farmer who was milking his cows. The man sat on his stool, his head pressed against the warm, soft flank of the animal, while Thoreau asked questions in the cold barn. Suddenly, and for no apparent reason, an ox standing near the two men "half lay, half fell, down on the hard and filthy floor, extending its legs helplessly to one side in a mechanical manner while its head was uncomfortably held between the stanchions as in a pillory."

Something about the sight moved Thoreau. It recalled to him the continuous discord over slavery that filled the daily newspapers. As he later wrote that evening, "The man's fellow-laborer the ox, tired with his day's work, is compelled to take his rest, like the most wretched slave or culprit."

The next day he wrote nothing in his journal. A fire of another kind had swept through Concord. That night he never even made it home.

Part III

Adaptations

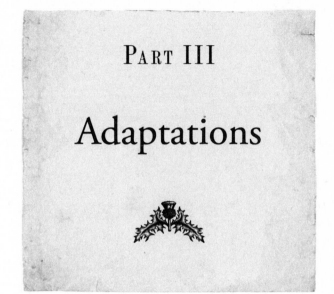

16

Discord in Concord

Later Franklin Sanborn would recall that he had been sitting at his desk in his carpet slippers when he heard the knock at the door. It was his custom to spend the evenings writing letters, preparing for the next day's classes, reading. Now and then he picked up a book from the stack on his desk, turned a page or two, then returned to his tasks. He might not have even heard the knock at first.

That he managed any sort of intellectual life that spring is remarkable. He lived in a state of constant anxiety. Again and again he imagined testifying before the Mason committee or being arrested for treason. He struggled to face the world with his characteristic insouciance.

Twice he had been summoned to appear before the congressional committee about his involvement in the John Brown affair. Twice he had fled to Canada. The threat of arrest weighed constantly. Yet every week he strode through town and up the long walk that led to the Old Manse, a glowering clapboard hulk built five years before the first shots of the American Revolution, an inauspicious residence made famous by Hawthorne in 1846 with his story collection *Mosses from the Old Manse.* Emerson had lived there in the 1830s and had written his first work, *Nature,* in the second-floor bedroom that overlooked the battlefield where the

Revolution had begun. (Hawthorne's desk had faced a wall; he did not care for the outside world to intrude upon his dark fantasies.) The Manse's current resident was Sarah Alden Ripley, a brilliant autodidact related to the Emerson clan, fluent in Greek and Latin. Ripley informally tutored many of Concord's most prominent young scholars, and Sanborn visited her weekly to read Theocritus in the original Greek.

He admired the woman's sly wit and deep learning. Her white hair wrapped in a lace bonnet, her gray eyes variously fierce and playful, she was sometimes capable of a despairing skepticism that blazed forth at unexpected times. It seemed to her that "the dirty planet on which we creep, if [it] were blotted out from existance [sic] would not be missed." More typically she was brisk and lively, a woman who read everything, consuming the era's most important works—often before either Emerson and Thoreau, who respected her opinions on literary, scientific, and political matters and solicited her advice on what to read next.

Ripley may have encouraged Sanborn's interest in the English Civil War that spring. Throughout March and April he dipped into a tottering stack of books on the religious and political conflict that had ripped England apart some two centuries earlier. As he reported to Theodore Parker, Clement Walker's *History of Independency,* an account of the vicious war between the Cavaliers and the Roundheads, had utterly engrossed him, offering a striking parallel to the present divisive moment in America. Civil war seemed increasingly likely that spring, especially as the nation launched into its quadrennial orgy of campaign promises and party demonization that occurred with each presidential election. This year the contest centered more than ever upon the issue of slavery and the territorial limits of the Cotton Kingdom. William E. Seward, the New York senator widely expected to be the next Republican presidential nominee, had described the situation a year and a half earlier in a speech entitled "The Irrepressible Conflict," which argued that "increase of population, which is filling the States out to their very borders, together with a new and extended network of railroads and . . . an internal commerce which daily becomes more intimate, is rapidly bringing the States into a higher and more perfect social unity or consolidation." If the states were evolving into something "higher" and more complex, their development nevertheless was threatened. Divisions between two incompatible societies were growing, and "these antagonistic systems are continually coming into closer contact, and collision results."

The press loved Seward. In a March review of his speaking style, the *New York Post* wrote that he addressed his topics "as coolly as Macaulay" and that his conclusions were as "thorough-going and trenchant . . . as we might suppose Agassiz . . . to make a reply to the rival theory of Darwin on the origin of species." But Seward's carefully laid plans to become the Republican presidential nominee had been upset by John Brown's attack the previous December. The failed insurrection at Harpers Ferry led many opinion makers and politicians to consider Seward's antislavery rhetoric too extreme, too alienating for the moderate Southern voters needed for presidential success. And this perception had opened the door for a long-shot candidate from Illinois.

Like many of his friends, Sanborn remained skeptical of the unlikely figure now appearing on the national political scene—a shambling rustic from Springfield, Illinois, visibly uncomfortable in his wrinkled, ill-fitting suit. Abraham Lincoln had been invited to deliver a speech in late February at New York's Cooper Union, which stood a block away from Charles Loring Brace's Children's Aid Society. A Westerner from humble origins, a melancholy autodidact, Lincoln's provincial manners concealed a shrewd strategist who grasped that the only way to win the general election was to exploit that razor-thin territory between pro-slavery apologia and radical abolitionism.

Lincoln felt with unquenchable conviction that slavery was wrong. His great adversary, Stephen Douglas, was fond of saying, "When the struggle is between the white man and the Negro, I am for the white man; when it is between the Negro and the crocodile, I am for the Negro." In 1859 Lincoln rebutted this racist humbug. Douglas was "blowing out the moral lights around us," he argued, "teaching that the negro is no longer a man but a brute; that the Declaration has nothing to do with him; that he ranks with the crocodile and the reptile; that man, with body and soul, is a matter of dollars and cents." But in the early months of 1860 Lincoln was cultivating an image of tolerant moderation. Addressing an imaginary audience of Southern Democrats in his public lectures, he tried to deflate the incendiary rhetoric that portrayed Northerners and Southerners as different peoples, emphasizing commonality instead. "We mean to remember that you are as good as we are; that there is no difference between us other than the difference of circumstances. We mean to recognize and bear in mind always that you have as good hearts in your bosoms as other people, or as we claim to have, and treat you accordingly." He also sought to defuse the hysteria prompted by John Brown's raid. "Old John Brown has just been executed for treason against the state," he told a crowd in Leavenworth, Kansas, in December 1859. Lincoln agreed with Brown's antislavery beliefs, but he also considered the man's hanging justified. No principle, however righteous, could "excuse violence, bloodshed, and treason." And for this very reason, he took a hard line toward those politicians below the Mason-Dixon line who discussed secession. If the South undertook "to destroy the Union," he warned, "it will be our duty to deal with you as old John Brown has been dealt with." At Cooper Union, he again addressed an imaginary crowd of Southern Democrats, accusing, "You charge that we stir up insurrections among your slaves. We deny it; and what is your proof? Harper's Ferry! John Brown!! John Brown was no Republican; and you have failed to implicate a single Republican in his Harper's Ferry enterprise."

That spring Sanborn supplemented Walker's history of the English Civil War with a series of others from the seventeenth century, including one by the parliamentary diarist Lawrence Whitacre and another by the royalist Edward Hyde, Lord Clarendon. Sanborn sided with the Puritans, of course; they reminded him of the abolitionists. In his mind, John Brown was the nineteenth century's Cromwell, its fierce and relentless leader against a Southern aristocracy that disregarded the will of the people.

But he was interested in another book as well—the same one in which

Sarah Alden Ripley was currently immersed. Possibly Asa Gray told her about *On the Origin of Species;* they were dear friends, and she frequently wrote him to share botanical knowledge. More likely, Thoreau encouraged her to read the new book. Later in the year, at a Thanksgiving dinner hosted by the Emerson family, she and the chemist Charles Jackson (Lidian Emerson's brother) discussed whether Darwin's theory was truly original or merely a restatement of older notions of metamorphosis.

Ripley also discussed Darwin's book during Sanborn's weekly visits to the Old Manse. Characteristically, he was interested less in Darwin's ideas than in the cultural politics of the book. That spring he wrote to Theodore Parker, telling him that Sarah Alden Ripley "has just been reading Darwin's book, . . . and likes it much as does Thoreau." Ripley may have either read the same copy that Thoreau had, borrowing it from the Concord Library, or been provided with a copy of her own by Sanborn. According to her son-in-law, James B. Thayer, "Mr. Sanborn, Mr. Channing, and other friends kept her largely supplied with the new books, and she read them eagerly, especially some of the newer contributions to natural science: the writings of Darwin and his supporters she cordially welcomed."

At any rate, Sanborn was not reading Darwin on the evening of April 3, when, as he later recalled in his autobiography, he sat at his desk in his carpet slippers, writing letters. It was not particularly late, but his sister Sarah was down the hall in her room. The Irish servant, Julia Leary, had already gone to bed. Now and then Sanborn picked up a book.

Then he heard a knock at the front door.

He put down his book and listened a moment. The knock sounded again, and he rose from his desk and went downstairs, a candle in his hand.

A man stood on the porch, his face nearly invisible in the dark. According to Sanborn's recollections, the man entered the house and asked, "Does Mr. Sanborn live here?"

Sanborn stiffly put out his hand. "That is my name, sir."

"Here is a paper for you."

Sanborn took the paper and stepped backward into the dim hallway.

The note was dated April 1, two days earlier, and it had come from the nearby town of Saugus: "*Sir:—*," it read,

> *The bearer, a worthy young man, solicits*
> *your aid in procuring employment.*
>
> *Buffum*

When Sanborn looked up, three more men were standing in his hallway.

"I arrest you," one of them said.

"By what authority?" Sanborn demanded.

"I am from the U.S. Marshal's office."

"What is your authority—your warrant?"

One of the other men spoke up. "We have a warrant."

Sanborn asked them to produce the document, to read it out loud, which one of the men began to do in a droning voice. The spell was broken by a noise from above. It was Sarah Sanborn, hurrying down the stairs, her eyes wide with fright. While the men stood in the hallway, she ran to the other door and began screaming as loud as she could, "Five men are arresting my brother!" (She had miscounted.)

Instantly the warrant was folded, and another man stepped forward to snap handcuffs on Sanborn's bony wrists. Then the four men lifted him off his feet.

Sanborn kicked his long legs. He twisted and jerked, his clothes tangling around his body. Within seconds he and the group of men were breathing hard and perspiring. The four struggled to get Sanborn out the door, but he kicked even harder. He yelled and spat and wrenched his shoulders, trying to get free. Then he planted his slippered feet on either doorjamb. The men swore and grabbed his thrashing legs, working to pry him from the door. Once they got him onto the veranda, he did the same thing, bracing himself against the posts of the porch, writhing as they carried him across the gravel walk, stiffening his legs on the stone gateway.

The men paused to catch their breath.

A carriage was waiting on Sudbury Road, and the group lifted Sanborn off his feet and approached it. He smashed its door. According to Edward Emerson, who was one of the first to arrive at the scene, Sanborn, "encompassed by a throng of men," was by this point "hoarse with passion." He continued "struggling convulsively" as the street began to fill with people. One of the men grabbed both of Sanborn's feet and tried to wedge him into the vehicle, but by this time the fire alarm—the bells of the First Church—was tolling, and a crowd had gathered outside Sanborn's house. Sarah ran into the street and grabbed the beard of the man holding her brother's feet. He let go.

At some point old Colonel Whiting, a carriage maker who lived on the corner of Main and Academy streets, ran up and tried to scare off the horses. His daughter, Ann Whiting, a prominent member of the Concord Ladies' Anti-Slavery Society, climbed into the driver's box and refused to budge. The crowd began to abuse the men who carried Sanborn. Some ventured to hit them. The lawyer J. S. Keyes ran into the street and asked if his client—he meant Sanborn—wished to petition for a writ of habeas corpus.

"By all means," Sanborn shouted, still struggling. Keyes then hurried to the house of Judge Hoar, who was already filling out the writ. John Moore, the town's deputy sheriff, rushed back to the scene and demanded the men surrender their prisoner. When they refused, he asked the crowd to help him free Sanborn. A group of Irish neighbors emerged from the dark and managed to extricate him from the four deputies. The neighbors stuffed them into the carriage and chased the vehicle, hooting and catcalling, as far as the nearby town of Lexington.

"We are having most exciting times here," Emerson's oldest daughter, Ellen, wrote to a friend the next day. "Have you ever enjoyed the interest of being awakened by alarm-bells and joining a street-fight, as most of the ladies and gentlemen of Concord did last night?" She added: "The town is in a high state of self-complacency, it flatters itself that this is the spirit of '76." Several houses away Louisa May Alcott wryly observed, "On Tuesday night we had a new sort of amusement called kidnapping. . . . I am so full of wrath I dont dare to unbottle myself for fear of the explosive consequences."

The fracas had lasted nearly two hours. When it was over, Sanborn recalled, he "was committed to the custody of Captain George L. Prescott . . . and spent the night in his house not far from the Old Manse, armed, for my better defense, with a six-shooter, which Mr. Bull, the inventor of the Concord grape . . . insisted I should take." This was the evening Thoreau didn't go home. Instead he spent the night at Sanborn's house, guarding it in case the four men returned.

Abolitionists across the North considered Sanborn's attempted arrest a call to action. According to the antislavery newspaper the *Independent,* the arrest was an abrogation of human rights. "Where are we? Whither are we tending? Have we a Star Chamber at Washington?" the newspaper demanded. "Are we under a Charles or a James?" The incident was the latest evidence of a political system so corrupted by slavery that "an inoffensive citizen has been rudely seized in his own house by a gang of armed men professing to act in the name of the Senate, manacled and dragged without hat or boots into the night air, to be smuggled into a carriage and hurried away to Washington without the possibility of defense." But for the prompt and courageous action of Sarah Sanborn, the paper continued, "the amiable and scholarly Mr. Sanborn would have been a prisoner."

The next day Sanborn was taken to the Boston Court House, where Judge Lemuel Shaw, the chief justice of the Massachusetts Supreme Court (and Herman Melville's father-in-law), listened to his case. Among those who attended the hearing were the abolitionist Wendell Phillips and the poet Walt Whitman, who had come from New York to discuss the publication of the third edition of *Leaves of Grass.* (It was here, perhaps, that Whitman first met Sanborn. In later years he liked to say, "I always hold Sanborn, Frank Sanborn, to be a true friend—to stand with those who wish me well.") Meanwhile the state legislature debated the legality of Concord's unified resistance. Some congressmen rejoiced that "the issue was met at Concord yesterday by the Democracy of that town, as a similar issue against tyranny was met in 1775." Others insisted that the case required no "particular sympathy, such as might perhaps naturally be excited in the case of a fugitive slave." Resisting arrest, after all, showed precious little regard for the rule of law.

At four o'clock in the afternoon, Sanborn was discharged from custody and escorted to East Cambridge, then returned by train to Concord. There Emerson, Thoreau, Bronson Alcott, and Thomas Wentworth Higginson, who had come down from Worcester, gathered before the Town House and vowed to protect him "against any Senate's office."

Thoreau was the first to speak that day. He told the assembled crowd that he had "heard the bells ringing last night, as he supposed for fire," but "it proved to be the hottest fire he ever witnessed in Concord." When the laughter subsided, he denounced "the mean and sneaking method the United States officials took to accomplish their purpose." There were some in the press who congratulated the town for conducting its defense of Sanborn in a lawful, orderly fashion. But Thoreau disagreed. "No," he said, "the Concord people didn't ring the fire alarm bells according to law—they didn't cheer according to law—they didn't groan according to law." Here he paused before a boisterous eruption of applause and said that as he didn't talk according to law, he thought he would stop and give way to some other speaker.

Sanborn appeared next, smiling at the "storm of applause, accompanied with a few hisses" that greeted him. When the crowd quieted, he thanked them for the "prompt, generous and unexpected manner in which they had come to assistance" the previous night. Then he held up a pair of handcuffs and rattled them over his head. "What are these an emblem of?" he asked.

Someone in the crowd yelled out, "Tyranny," but Sanborn corrected the person:

"It is the badge of *slavery*."

The yard before the Town Hall filled with more cheers and applause, and Sanborn shook and rattled the manacles with greater force, his voice rising with indignation. The Southerners held the entire country as "much in bondage as their own slaves." Slavery had implicated everyone in the nation, regardless of where they lived or what they believed. If anyone dared strike out in freedom, "the whole power of the national government is brought to bear to crush him."

The only response, therefore, was war—not the kind of war declared by governments but a very personal sort: a struggle to extinguish one's foes. According to a reporter for the *New York Herald,* "Mr. Sanborn continued, by saying that if those ruffians who attempted to carry him off last night had been killed in the act, the deed would not have been deemed unlawful. . . . They ought to have been killed. (*Applause.*) Such men are not killed. They die like vermin. (*Renewed applause*). . . . When the Senate, or the House of Representatives, or the President, act under the mandates of the slave owners, they must be resisted by every way and by every means. (*More applause.*) Mr. Sanborn said he was ready to meet any other offence against the South in the same manner he had done last night. Dealing with the Southerners was not dealing with men, it was dealing with demons. The system of slavery should be opposed with force. . . . There is no law to protect tigers and hyenas—and they must be met and dealt with as such."

Sanborn had learned quite a lot in the past twenty-four hours, he told the crowd. He had learned that John Brown was right, that the slave interest could be destroyed only by violence. And he prayed that God might help him now "in pursuing the same course."

These last words brought about the loudest cheers of all. The area before Town Hall was filled with women holding parasols, men in frock coats and top hats. They shouted and cajoled, clapped their hands. They jeered every time the South was mentioned. At that moment Franklin Sanborn must have felt as though he had lived up to the spirit of Concord. He had enacted Thoreau's civil disobedience and assumed the role of Emerson's self-reliant man. Most important, he had vindicated old John Brown. When the applause died down, he held up his manacles and rattled them again, inspiring fresh waves of cheering.

The Sanborn incident inflamed the nation, bringing it one step closer to dissolution. Samuel May, Bronson Alcott's brother-in-law, reflected the opinion of many antislavery campaigners when he complained, "What we are coming to in our country seems to me evil—fearfully evil. . . . The attempt to kidnap Sanborn shows plainly enough that the pro slavery party means to stick at nothing."

Like a ripple in a pond, the event spread well beyond New England's borders, appearing in newspapers from Bangor to New Orleans, casting a pall over the Democratic Party's national convention, held three weeks later in Charleston, South Carolina. The party's front-runner, Stephen Douglas, the man who had beaten Lincoln two years earlier in the Illinois race for Senate, urged calm among his fellow party members. Sectional harmony was surely necessary for the preservation of the Union, he argued. But peace with the North had become anathema to the secessionist wing of the party. John Brown was no freedom fighter, these Southern Democrats argued, but a lawless renegade intent on plundering private property. "Ours is the property invaded," declared William Yancey of Alabama, "ours are the institutions which are at stake; ours is the peace that is to be destroyed. . . . Bear with us, then, if we stand sternly upon what is yet that dormant volcano, and say we yield no position here until we are convinced we are wrong." Yancey and other secessionists believed that Sanborn was merely behaving like any other abolitionist, thumbing his nose at the Constitution while ascribing his illegal behavior to a higher purpose. He and his allies demanded that the Democratic Party add a plank to its platform establishing a "slave code" to protect slavery in the territories. Douglas refused this demand, and on April 29, when it became clear that Northern and Southern Democrats could not agree, the Southerners walked out of the convention, splintering the party and opening the path for a Republican nominee to win the presidency.

17

Moods

Not long after the Democratic Convention's fractured conclusion, Louisa May Alcott once again fell into the vortex—into that fever of language and plot that seemed more urgent than anything in her day-to-day life. She followed the nation's political crisis with interest, proudly noting Concord's resistance in the Sanborn affair and disdaining of the Democratic Party and its defense of slavery. But nothing seemed real to her until she put it into words. Nothing made sense unless ordered on the page. Her first novel based its principal male characters on Emerson and Thoreau, casting both men as romantic suitors to the central heroine. But the work was also a morality tale about North and South, about the issues that concerned most Americans in the summer of 1860, its plot an effort to exorcise the irresolvable conflicts the nation was experiencing.

The precipitating event was her older sister Anna's marriage. Anna wed John Pratt on a day when the apple blossoms looked like snow outside Orchard House. Louisa's abolitionist uncle Samuel May performed the service, and Thoreau, Sanborn, and Emerson were in attendance. When the ceremony was over, the small gathering joined hands and danced around the young couple. Emerson kissed the bride, and the day was "all grace and becomingness," as Bronson recorded in his journal. Louisa felt otherwise. "I mourn the loss of my Nan," she wrote on the same day, "and am not comforted."

More disruptions followed. The same week as the wedding, Bronson and Abba Alcott hosted a reception for John Brown's widow, who had come all the way from North Elba, New York, to visit her two youngest daughters at Sanborn's academy. Mary Brown arrived in Concord with her widowed daughter-in-law, whose husband, Oliver Brown, had fired the first shot at Harpers Ferry, killing an innocent black baggage handler. Oliver had himself been killed in the small engine house where Brown and his men retreated. The two women brought with them a fatherless infant, Oliver's son, whom Louisa cuddled and held during much of the visit. She found herself drawn to the widowed Mary Brown, "a tall, stout woman, plain, but with a strong, good face, and a natural dignity that showed she was something better than a 'lady,' though she *did* drink out of her saucer and used the plainest speech." For Louisa, tragedy had ennobled the woman, rendering her into something statuesque and even regal.

She had already begun to think about writing a novel when, in late June, the Hawthorne family returned to Concord, moving next door into a colonial-era house the Alcotts had inhabited back in the 1840s. "Mr. H. is as queer as ever," Louisa soon wrote a friend, "and we catch glimpses of a dark mysterious looking man in a big hat and red slippers darting over the hills or skimming by as if he expected the house of Alcott were about to rush out and clutch him." In truth, she was fascinated by the novelist, whose self-styled romances had made him one of the most respected writers in the country. It was her wish to write a book as serious and beautiful as one of his, a book that would explore the deepest recesses of the human heart. But as ever, the need for money and the demands of the marketplace kept her busy churning out potboilers and overheated love stories. On June 29 she attended a welcoming party for the Hawthorne family, which included the novelist's diminutive and sickly wife Sophia and their three children, Una, Julian, and Rose. The Hawthornes had spent the past decade in England and Italy, and they carried with them now a curious combination of worldliness and clannish insularity seldom seen in Concord. Strawberries and cream were served at the Emerson house, where the party was held, and Thoreau and Sanborn were once again in attendance.

Louisa almost certainly read Hawthorne's latest novel that summer.

The Marble Faun describes a trio of American artists who collectively lose their innocence in Europe. Kenyon is a rational, sunny-tempered sculptor in love with Hilda, a fair-haired painter who copies Italian masterpieces. As usual with Hawthorne, the "fair" heroine is paired with a "dark" woman, Miriam, whose subversive erotic energies are manifested in her daring and original paintings. The three Americans encounter an Italian named Donatello, who is a kind of copy, too: he bears a striking resemblance to Praxiteles's sculptured faun. Hawthorne hints that Donatello is in fact the descendant of ancient fauns—those rustic forest gods who were half-goat and half-human—and he is presented as an Adam-like primitive, a missing link "standing betwixt man and animal, sympathizing with each" and content to enjoy "the warm, sensuous, earthy side of Nature." Despite his genial nature, there is something tempestuous about Donatello. "If you consider him well," Miriam remarks early in the novel, "you will observe an odd mixture of the bull-dog, or some other equally fierce brute, in our friend's composition; a trait of savageness hardly to be expected." When Donatello falls in love with Miriam and eventually commits a murder on her behalf, every character is permanently altered.

Several decades after *The Marble Faun*'s publication in 1860, the critic Frank Preston Stearns would comment, "It is a wonderful coincidence that almost in the same months that Hawthorne was writing this romance, Charles Darwin was also finishing his work on the 'Origin of Species.'" Stearns was the son of George Luther Stearns, a member of the Secret Six, and he believed that taken together, the two books perfectly reflected their historical moment. "Hawthorne did not read scientific and philosophical books, but he may have heard something of Darwin's undertaking in England," he speculated. If the "skeleton of a prehistoric man discovered in the Neanderthal cave, which was supposed to have proved the Darwinian theory, does not suggest a figure similar to the 'Faun' of Praxiteles," Stearns continued, ". . . the followers of Darwin have frequently adverted to the Hellenic traditions of fauns and satyrs in support of their theory."

For Stearns, however, Hawthorne's novel was a bracing counterweight to the amoral determinism he detected in the theory of evolution. Hawthorne had "made a long stride beyond Darwin, for he has endeavored to reconcile [an evolutionary] view of creation with the Mosaic cosmogony; and it must be admitted that he has been fairly successful." Stearns believed that Hawthorne's novel described a different sort of evolution than Darwin's book. It portrayed human progress as an endless struggle between moral advance and worldly sin. Though it was typical of Hawthorne's writing that sin was often the more compelling subject, the moral advance invariably won out. Each of Hawthorne's four principal characters experiences a fall from innocence that is "really a rise in life," Stearns wrote. "Do not all judicious parents protect their children from a knowledge of the world's wickedness, so long as it is possible to prevent it,—and yet not too long, for then they would become unfitted for their struggle with the world, and in order to avoid the pitfalls of mature life they must know where the pitfalls are."

A rumor arose almost immediately among New England's literati that Donatello was based upon Henry David Thoreau. Hawthorne had known and admired Thoreau for nearly two decades, and there are some obvious similarities between his fictional character and the Concord naturalist. Both play the flute, both share a fondness for the woods and possess an uncanny ability to tame small animals. More important, both seem to embody an Arcadian past that stands in sharp contrast to a corrupt and sophisticated society. It was Hawthorne's special insight to see something both deeply attractive and potentially dangerous about Thoreau's rusticism, his willful embrace of the primitive. The novelist believed there was something ultimately illusory about the attempt to live entirely alone in the wilderness, separated from humanity. Donatello's bestial moods and instinctive actions, as well as his unabashed relish of the sensual world, repel Hilda for much the same reason they attract Miriam: by aligning himself with brute creation, Donatello jeopardizes his basic humanity.

Hawthorne's presence in Concord was an inspiration to Louisa, who began her novel shortly after the writer and his family arrived next door. "Genius burned so fiercely," she later recalled of this period, "that for four weeks I wrote all day and planned nearly all night, being quite possessed by my work. I was perfectly happy and seemed to have no wants." The result of this feverish period was *Moods,* the chronicle of a love triangle between an impetuous heroine and a pair of male characters. The plot is simple: a lively young woman named Sylvia Yule marries Geoffrey Moor, a refined intellectual based on Louisa's esteemed neighbor, Ralph Waldo Emerson. Sylvia admires Geoffrey and appreciates his bountiful library, but she is physically drawn to Adam Warwick, an earthy iconoclast who tames wild creatures and women alike. Adam's first and last names are significant: they refer not only to Thoreau's "preference for Adamhood," as Bronson had called it, but to a private war within the man. For Adam has a secret past: he is betrothed to a Cuban woman who may or may not possess African blood and who exerts a potent erotic charge of her own.

The novel is a nineteenth-century soap opera studded with miscommunications, crossed intentions, and ardent declarations of love. But it is also a product of its moment. Throughout the book, matrimony is equated with slavery; slavery is in turn portrayed as a bitter and failed marriage between North and South—a marriage from which neither party can escape. And in the final analysis, romantic love proves to be little more than a form of Darwinian competition.

Because she is impulsive and inexperienced—governed by *moods,* in the novel's parlance—Sylvia initially mistakes Geoffrey's kindness for love. Only after marrying him does she realize her error. "The husband's charm had lost its virtue," Alcott writes, "when the stronger power claimed her." That stronger power is physical desire, which burns through the novel from start to finish. Driven by "the old passionate longing" to be with Adam, Sylvia repeatedly "felt the rush of a coming impulse, knew that it would sweep her into Warwick's arms, there to forget her duty, to forfeit his respect."

Readers are prepared for this consuming passion, because Adam Warwick has already experienced something similar with Ottila, the jealous Cuban seductress he met while traveling through the Caribbean. The illicit relationship, which Adam describes as "a spiritual slavery," is a metaphor for the United States, especially its misbegotten marriage between

Northern industry and Southern chattel slavery. The "chill of an Arctic winter" in Warwick's "blood and brain" collides with Ottila's "climate of the passions," where "the pestilence of slavery" lingers like a cloying perfume in the tropical air. Warwick's uncontrollable passion for Ottila mirrors the relationship between Northern commerce and a slave-based economy that made industrial expansion possible.

Sylvia's entrance into Warwick's life does little to effect a resolution. She is, after all, the child of a Northern father and a Southern mother. From her father "she received pride, intellect, and will; from her mother passion [and] imagination. . . . These two masters ruled soul and body, warring against each other." Early in the novel Sylvia agrees to join a country excursion with Adam, Geoffrey, and her brother Mark. The four friends boat down a river before setting up camp near a farm, where they spend the night in the hayloft of a barn. Louisa takes great pains to describe the sleeping arrangements of her characters—the men on one side, Sylvia on the other—but this separation doesn't prevent her heroine from awakening with a "curious thrill . . . from deepest slumber," feeling "a hot tingle through blood and nerves" that causes her to "think . . . of fire."

The next day the flame of desire is literalized. As Sylvia and Adam wander through the woods together, pausing to observe a war between red and black ants (Louisa's nod to a similar incident in *Walden*), they come upon a forest fire much like the one Thoreau had recently visited outside Acton. "Many acres were burning," Louisa wrote, "the air was full of the rush and roar of the victorious element, the crash of trees that fell before it, and the shouts of men who fought it unavailingly." Feeling "mutinous" and "perverse," Sylvia disobeys Adam's instructions to remain in safety and plunges into a thick copse, where she is soon lost in the smoke-filled forest. On the verge of succumbing to the smoke and flames, she hears a sound, and Adam appears and rescues her.

Marriage plots invariably dramatize the conflict between personal desire and social respectability, and *Moods* has been taken as a statement of Louisa's doubts about the institution's ability to accommodate the disparate elements of erotic passion, companionate friendship, and female autonomy. Alcott's biographers have also seen the novel as a critique of her parents' unequal and sometimes unhappy union. More vivid than either of these concerns, however, is the book's meditation on physical passion—on the way desire overwhelms the self, consuming reason and propriety. Sylvia and Adam's sexual attraction is instinctual, a biological imperative hardwired and difficult to thwart. When she refuses to confess her attraction to Adam, he replies, "Shy thing! I will tame you yet, and draw you to me as confidingly as I drew the bird to hop into my hand and eat. You must not fear me, Sylvia, else I shall grow tyrannical; for I hate fear, and like to trample on whatever dares not fill its place bravely." The example he provides of something that fills "its place bravely" comes directly from Darwin, or more likely from Thoreau's discussions of Darwin—the "little mosses" crowding one another near a stream.

The difficulty for Louisa was in imagining how to integrate erotic desire into antebellum American society. She struggled to satisfy both the requirements of her plot and the expectations of her genteel audience, ultimately killing off Warwick in a shipwreck. In an act of self-sacrifice, Adam saves Geoffrey Moor as their ship plunges into a swirling vortex, much as the *Pequod* does at the end of *Moby-Dick*.

But this is not the way Louisa *wanted* her book to end. Alcott imagined Sylvia living in contented solitude, her personality enriched by its former passion. Friends and publishers found this conclusion alarming; it rewarded adulterous longing and female sensuality, and it thwarted the instructive moral justice that novels were supposed to dispense. In the end, Louisa capitulated. She inflicted a wasting illness upon her heroine. Sylvia dies in the arms of Moor, who continues to love her despite her final protestation that she cares for him as a friend and not a husband.

In many ways, *Moods* signals a more general upheaval in American literature. Like Rebecca Harding Davis's "Life in the Iron-Mills," published in 1862, *Moods* represents a shift away from transcendental idealism and domestic sentimentality. It forecasts a darker literary naturalism that we typically consider a by-product of the Civil War, which is often thought to have extinguished forever the lofty romanticism of the antebellum period. This shift can be located in the novel's key term, *moods*, which describes the capricious and uncontrollable actions of Alcott's heroine in terms that are more naturalistic than spiritual. Sylvia is driven less by reason and free will than by instinctual desire: by the physical, sexual, and biological. Like a wood chip on the ocean, she is guided by hidden currents over which she has no control. "Instinct," Louisa wrote of her character, "subtler than perception, prompted both act and aspect."

Such insights were uncommon in the domestic fiction of 1860. They belong in general to a later period, to the cultural milieu that inspired Kate Chopin's *The Awakening*, written in 1899 and chronicling another seemingly dauntless heroine trapped by social expectations, physical desire, and the longing for independence. The difference between the two novels is that Chopin possessed a ready explanation for the competing forces that ultimately destroy her character. Darwinian theories had been widely adopted during the two decades before her novel, making their appearance in works by Stephen Crane and Jack London, and in the same year as Chopin's novel, in Theodore Dreiser's *Sister Carrie*. All these narratives presented characters who were driven by large impersonal forces—biology, sexuality, economics—in a vision that owed a great deal to Darwin's naturalistic model of the world. These characters either adapted to their environment, like Carrie in New York, or died because they could not, like London's unnamed Alaskan who cannot light a fire during the bitter cold.

The Awakening's physician, Dr. Mandelet, explains the problem to Edna Pontellier, Chopin's heroine, after she witnesses the birth of her friend's baby. "The trouble," Mandelet observes, "is that youth is given up to [romantic] illusions. It seems to be a provision of Nature, a decoy to secure mothers for the race. And Nature takes no account of moral consequences, of arbitrary conditions which we create and which we feel obliged to maintain at any cost." Mandelet's point is that physical attraction is necessary for successful mating. Edna's culture mistakes fleeting attraction—biological instinct—for a lifelong attachment that may be necessary to ensure social order but has little basis in nature.

Adam Warwick seems to recognize the disruptive nature of sexual desire when he confesses just before drowning, "This peaceful mood may not last . . . but it brings that rare moment . . . when we seem to see temptation at our feet." Louisa herself experienced very little peace during that summer. She was too immersed in the tangled, fervid romance of her imagination. At some point she read her new novel to Bronson and Abba, who discussed the book with friends and neighbors. Before long she confided in her journal, "Mr. Emerson offered to read it when Mother told him it was 'Moods' and had one of his sayings for a motto."

Years later she would look back at this period of creation with wonder. As so often happened when she wrote, characters had assumed lives of their own, ignored her plot, and behaved independently. She had set out to write a romance about the problem of marriage, but slavery and biology and the determinism of inheritance had seeped into the structure of the book. Without meaning to, she had written an allegory about the North and the South, about the nation's mixed inheritance, its tangled and apparently insoluble history. She had depicted a world governed by impersonal forces and instinctive drives, by the struggle for dominance, a world in which self-determination is difficult if not entirely futile. Although she had no way of knowing it, she had written about the dangers that lay just ahead.

18

Meditations in a Garden

Increasingly that summer, Bronson Alcott had fallen into a vortex of his own. He listened with keen apprehension as Louisa read aloud her new novel, imagining Sylvia Yule and Adam Warwick with the same vivid attention with which he always pictured the hero in his favorite book, *Pilgrim's Progress*. Louisa's characters yearned for an end to their tumultuous emotions. They prayed for an uncertain salvation. They wanted the "restless, domineering devil that haunted" them to be "cast out" forever. Bronson could sympathize.

He sat in his garden, clutching his gold-headed cane, inhaling the fecund odors of squash and tomatoes. His was a monistic philosophy—predicated on a single unified reality—yet more immediate divisions seemed to be pulling this reality apart. Strife over slavery threatened to end the experiment in American democracy once and for all. A July editorial from Richmond, Virginia—a town Bronson had visited and loved as a young peddler of Yankee wares—discussed the schism in moral terms: "Resistance to wrong and injury—to tyranny, whether of one man or eighteen million—is the cherished birth-right of every citizen of the Federal Union." The tyranny being referred to was the nomination of Abraham Lincoln as the Republican Party's presidential candidate, an event recently accomplished at the national convention in Chicago. The editorial concluded by advocating disunion: the South's prompt and permanent exit from the United States of America.

Even Concord had become a cauldron of dissent. Some in town wished to maintain the fragile Union at all costs. Old Moses Prichard, stooped and spare, his face rucked with years spent outdoors in his garden, had long supported upholding the Fugitive Slave Act. His wife, Jane, a reader of French literature and a woman of vinegary wit who was now suffering from cancer, was that rarity in Concord: an outspoken opponent of the antislavery movement altogether. Alcott enjoyed speaking to both of them, mainly because he valued their advice about gardening. Husband and wife had been born the year George Washington assumed the presidency, and both could remember New England when it was little more than a loose collection of subsistence farms and villages, each with its clapboard church and family pews. Alcott increasingly found conversation with the Prichards tinged with melancholy. The nation they had grown old with and had loved with familial pride was afflicted now with madness and folly. The town's abolitionists had grown ever more strident, condemning all who declined to share their views. To the Prichards, this did not seem quite neighborly.

Others took their cue from the town's central figure, Emerson, who increasingly felt that any political relation with the slaveholding South was a relation with evil. Recently he had written his friend and fellow member of the Saturday Club, Oliver Wendell Holmes, to scold him for his continued support of the Union. "And for the Union with Slavery no manly person will suffer a day to go by without discrediting disintegrating & finally exploding it," Emerson stormed. "The 'union' they talk of is dead & rotten."

Sophia Thoreau—Henry's small, dark-eyed sister and one of the town's most passionate abolitionists—was in perfect agreement: "My motto has long been 'no union with slaveholders,'" she announced to anyone who would listen. Like others in town, she believed more active measures were required to break the yoke of slavery. It had once been enough to advocate "disunion"—in this case, the secession of New England from the Southern states—but the attempted arrest of young Franklin Sanborn proved that John Brown had been a prophet and a visionary. Tensions had simmered since the New World's founding, since that infamous day when an African slave was first brought to Virginia's shores. Those tensions were now about to boil over.

Alcott himself was divided. On the one hand, he revered the courage of Brown and believed in the cause of emancipation. On the other hand, he detested violence and discord. Over the years he had devised a theory of community that he called, with typical imprecision, Personalism. Alcott believed that all "souls are *alike* essentially in virtue of being One." Every person possessed a spirit that partook of a larger spirit. The Person referred to in Alcott's theory was God, and Personalism was his effort to counter the selfish excesses made on behalf of individualism, which too often emphasized the differences between people at the expense of commonality. "Individualism is brute, egotistic, inhuman," Alcott announced in his journal. The American emphasis on the self was divisive and unproductive; it could be overcome only when people realized that their souls were "similar and partakers of a common Personality in the One." Alcott was being "personal" when he spoke before schoolchildren and their parents or when he delivered his inimitable "Conversations" to baffled listeners. He was drawing upon the divine essence available to all people to create friendship and harmony. If the "personal" was the godly source of all creative power, it was also the basis of community: "We make progress only as we find that common basis and ground of agreement wherein all have their root."

Beneath the froth of Alcott's abstractions was a more or less coherent philosophy. Calvinism was wrong. People were not depraved but were capable of moral perfection. The problem was society: its constraining institutions, its narrow traditions, its tendency to promote selfish individualism. Alcott had tried to combat society's evils nearly two decades earlier, when he founded an experimental utopian community called Fruitlands. His dream had been to create a self-sufficient society joined in comity and a mutual striving toward perfection. Fruitlands was to be a seed from heaven planted in the yielding earth of New England. When the project collapsed in indigence and acrimony, he had tumbled into a slough of despond so murky no light could penetrate. He refused to eat or speak. Something similar had happened in the 1830s, when he was forced to shutter the Temple School. He had been struck by lethargy, sickened in the soul.

These days he preferred whenever possible to float through life—to drift along life's sunny currents, as buoyant as the water lilies his friend Thoreau loved to collect. He sat in his garden and allowed his mind to wander. As he would later record in his journal, the day was hot, even for the last day of July, but his corn and beans were thriving after yesterday's rain, and he had just finished sowing turnip seeds for the autumn. Inside the house, Abba was busy with domestic chores. May was likely painting, and Louisa was still in the grip of her novel. Alcott thought he understood the thrall and challenge experienced by his temperamental daughter as she tried to wrestle vision into words. Many years ago he had seen spirits and heard the voice of God. The days of his youth had been lit with portent— with a conviction that spirit suffused everything. "Our thoughts are the offspring of that divine power," he wrote a friend during this time, "which, when freed from the obstacles of human authority, and the influences of human circumstances, feels itself the agent of its own advancement, and the source of truth and virtue." American society might writhe and stumble in its endless search for material gain. Politicians might pervert the intention of the Declaration of Independence. But *truth* remained stable and fixed. "Truth and love are alone immutable and immortal," he wrote, "the unchanging elements of the divine economy."

Earlier that morning he had stumbled across a brief but suggestive item in the local newspaper and had promptly pasted it into his journal. Entitled "THE PROGRESS OF MANKIND," the squib began by mentioning a Swedish poet and feminist named Karl Jonas Ludwig Almquist who believed "the universal striving of humanity on earth has been to knit various races together." Alcott could only tentatively agree. All races were certainly connected by the divine soul, he thought. All races deserved to be treated equally. He was less certain about their becoming knit together.

But that wasn't why he cut out the piece. He was interested in the speculations of its anonymous author, who stated that Almquist believed the human race "constitutes the ideal principle of the planet, its soul, its spiritual life." This raised a number of questions for the newspaper writer. Had the world, for instance, "reached such a point of development, such a lofty height . . . in its energies of culture," that it might be regarded as mature? Or was it possible it had "not yet attained its highest culmination"? Were the Earth and its human occupants to "be regarded as old or young"? Was humanity now poised to slide once again into savagery and barbarism—questions that "anxiously occupy many of the profoundest thinkers on the destiny of mankind"?

The issue gnawed at Alcott. He worried that human culture *was* in fact regressing. He feared that America was sliding into dissolution. The civic order, the social pact constructed by his grandparents' generation, was breaking under daily assaults. "We belong now to a Church with one member only," he told whoever would listen. This emphasis on the individual was the root cause of slavery; nothing but sinful selfishness could explain the willingness to fetter the body and soul of another human being for dollars. When we "at last get free of our individuality, and become persons indeed," then we might "partake of that which unites and relates us to one another."

His thoughts were interrupted by the sound of visitors. Like a stork in a frock coat, Emerson suddenly appeared from the apple orchard. That summer he frequently stopped by the Alcott home, admiring the progress of Bronson's ambitious garden. In part, this was Emerson's way of preventing his friend from brooding too much. But he also took childish joy in Alcott's vegetables and herbs, which invariably surpassed his own feeble efforts in the garden.

Today he was accompanied by a visitor. Tall and pale, his face shadowed by an incipient beard, Moncure Daniel Conway was another young acolyte of Emerson's. In his short life he had been a Unitarian minister, a prolific journalist, and a tireless advocate of women's rights and abolitionism. Now he had a new enthusiasm. Living in the raw but rapidly developing state of Ohio, Conway was among the first in the West to have acquired a copy of the *Origin of Species*. He had absorbed it as quickly as he had once absorbed Emerson's philosophy, accepting it with the same ease and readiness. The book was colossal, he thought. It was epoch-making. By July 1860 he had already begun to introduce evolutionary theory into his sermons: "Now comes Darwin and establishes the fact that Nature is all miracle, but without the special ones desired," he told his Ohio congregation: "that by perfect laws the lower species were trained to the next higher."

Conway greeted Alcott by handing him a copy of the *Dial*. The original *Dial* had been founded two decades earlier by Emerson and the brilliant critic and feminist, Margaret Fuller, who had died ten years before in a tragic shipwreck off the coast of Long Island. Alcott had provided the journal with its name, and a few of his "Orphic Sayings" appeared in its early numbers. Recently Conway had decided to revive it. Literary Concord followed with interest the progress of the new periodical, which carried a hodgepodge of poetry, serialized novels, and essays about slavery, ghosts, feminism, and spontaneous generation. Sanborn, always jealous of Emerson's other young followers, had pronounced his opinion earlier that spring: "Conway's Dial has some good things in it, but is on the whole dull and needless."

Each issue contained book reviews, including ones on the Lincoln-Douglas debates, on Whitman's recently revised edition of *Leaves of Grass,* and on *Adela, or the Octoroon,* an antislavery novel hastily written by Hezekiah Lord Hosmer to cash in on the "tragic mulatta" craze. The March issue also carried one of the earliest American reviews of Darwin's *Origin*. Almost certainly written by Conway, it compared Darwin to "Newton pondering the falling apple, Young, the aerial soap-bubble, Goethe the bleached skull of a ram, deducing from these central laws." Darwin's central law provided a glimpse into the beginning of time. And unlike the theory of gravity, its scientific principles could be translated into social terms. "The awkward, the monstrous, the slow, the useless" became extinct, while "the free, the fleet, the valuable . . . escaped and perpetuated themselves." This was a political as well as a theological statement. It suggested the eminence of America among nations, and it implicitly criticized the practice of slavery. Conway discerned in natural selection a divine intention: the desire to propagate the "free" and the "valuable" at the expense of the "monstrous" and the "useless." These last two words were code for the Southern plantocracy—a class of indolent usurpers who were rapidly becoming obsolete. The review closed by calling Darwin's book a "timely and excellent work, which brings with it inevitably the crisis of inquiry into this much discussed question of the origin of species."

The current issue Conway handed Alcott contained another review of Darwin's book, this one different—and more defensive—than the first. It was written by someone identified only by the initials M.B.B. No true objection could be made to Darwin's theory, the review began, because "however profound our ignorance, . . . system and order lie at the foundation of all, as the action of the Creator's will." If Conway accepted Darwin's theory as evidence of progressive development, M.B.B. sought to combat its central defect, which was that "it supposes that everything which does not serve a material purpose, is subject only to chance or accident; or, perhaps, that there is nothing existing but that which is of material use."

One had only to study nature to realize that this was not true. God was an artist who cared as much for the beauty of His creation as for the survival of a particular species. "The highest types of beauty most often combine with forms least able to withstand the fierce struggle for existence," wrote M.B.B. Why would God bother to produce such infinitely beautiful creatures if brute endurance were the supreme goal? Across the planet God's "creative skill seems as manifest in the production of qualities beautiful, as in those of simply material use. 'Nature puts some kind of pleasure,' says Thoreau, 'before every fruit; not simply a calix behind it.'"

In the end, M.B.B. refused to consider that life might be the product of material causes alone, noting that "Mr. Darwin's theory leaves a God in the material world; for here we see the prevalence of *law*." Gravity was a law that kept planets in their orbit around the sun. Light waves traveled faster than sound waves. These principles suggested a supreme Lawgiver, a sentient Being, for "'what is fixed as such requires and presupposes an intelligent agent to render it so.'"

Alcott could not help but agree. Darwin's ideas stripped life of its grandeur. They made a mockery of one's deepest certainties that the world had been wisely designed. Alcott could not imagine a universe so perversely cruel as to produce life without meaning. Nor could he bear to live in a world that was reduced to the most tangible and daily phenomena, to random change and process. Materialism placed one at the very rim of the world, forced one to confront the possibility of chaos and void. It stripped one of purpose—of a purpose that seemed manifest in all nature. Who could witness the gorgeous ruddy sunrise of a summer morning and not see behind it God's joy in His creation? Who did not feel as if God were in the world and within each person?

The reviewer for the *Dial* made a similar point. Look at the human eyes, he wrote. "How shall we account . . . for the shaping of these organs, which *natural selection* could have made only for use in preserving the life of the race . . . ?" How could the struggle to survive explain "a form moulded to such perfection of beauty as that which the artist has copied in the statue of Apollo?" The eye was not merely functional. It was a marvel of optics *and* an aesthetic object—endowed with beauty that served no other purpose than to please those who stared into its depths and encountered the soul. If natural selection alone accounted for human development, such gorgeous ornaments would be entirely unnecessary. "I think natural selection would give us nothing but Calibans," declared M.B.B., referring to the bestial, subhuman character in Shakespeare's *The Tempest*. And this allusion led to a statement of the *Dial*'s abolitionist policy. "As it is, however, nothing but long continued degradation and oppression suffice to even partially efface the image of God in the human form."

Alcott kept the copy of the *Dial* as Emerson and Conway took their leave. He remained in the garden awhile longer, spading the soil, bending down to pluck weeds from his rows of tomatoes. Emerson had invited him to continue their conversation after dinner, and several hours later, having flipped through the *Dial,* he walked to the tall white house in the summer twilight.

The Emerson house was lively that evening, children coming in and out, conversation in the parlor. Conway wanted to discuss Darwin's theory, which he believed was in perfect agreement with transcendentalism. "In the year (1836) when Darwin abandoned theology to study nature," he would later write, "Emerson, having also abandoned theology, published his first book, 'Nature,' whose theme is Evolution." Conway recalled reading Emerson's statement that "the way upward from the invisible protoplasm to the highest organism—gave the poetic key to natural science." The statement, he now thought, had been prophetic, inspired. Science and revelation had fused in Emerson's insight, suggesting deep laws undergirding the universe. Later that year, in the October *Dial,* Conway would complain that religion had yet to absorb the new theory. "Our popular Christianity has not fulfilled the law of the higher formation. It must everywhere sum up all the preceding formations, and lose none of their contributions, as the animal generations are summed up in the forehead of man."

Emerson and Conway discussed these things until late in the evening. Alcott objected to their enthusiastic acceptance of Darwin's theory, but to no avail. For once in his life, he felt outside the current of new ideas. That night, when he returned home to his study (Louisa was still upstairs, still writing furiously), he recorded in his journal: "Evening: I am at Emerson's and talk on the Darwin Book and [its] Principles largely with him and Conway."

It was the last time he would mention Darwin that year.

19

The Succession of Forest Trees

Henry Thoreau's discourse before the Middlesex [County] Agricultural Society" was a resounding success, Bronson Alcott wrote in his journal in late September. The society's fair was held annually in Concord, drawing enormous throngs of people from neighboring towns and villages. Because this was high-minded New England, the event combined self-improvement with agricultural contests and exhibits. Alcott and Emerson had both delivered lectures in the recent past, their inspirational messages punctuated by the plaintive lowing of penned cattle and sheep. Cash prizes ranging from three to ten dollars were awarded for "best porkers," "best geese," and "best stallion." There were countless other honors for finest apples, best-made boots, sweetest peaches and plums, best watermelon, butter, bread, and flowers. John Brown's daughter, Sarah, received a dollar award for her exquisite needlework.

That year the fair took place "under rather unfavorable auspices," because of "the very inclement state of the weather." The skies were purple with storm clouds, and a silvery rain slanted off and on, dousing the fairgrounds. In an effort to stay dry, people crowded into the exhibition hall, which was "ornamented by suspending carpeting from the upper part of the building." By two o'clock the weather had cleared enough to allow the marching band to escort a crowd down Main Street and into the Town Hall, where Thoreau delivered what became his most widely read piece of writing during his lifetime. "The Succession of Forest Trees" was soon reprinted in newspapers across the country, first by his acquaintance Horace Greeley at the *New-York Tribune,* and then by many other smaller periodicals. The essay was the result of Thoreau's encounter with Darwin.

Despite its evolutionary overtones, Alcott enjoyed his friend's talk "on Nature's Methods of planting forest trees by animals and winds," finding it "admirable and interesting" and every bit as entertaining as "Wild Apples." The lecture proposed to explain why oak forests were replaced by pine forests when cut down and vice versa—a phenomenon Emerson, Horace Greeley, and many other reasonably informed observers considered an impenetrable mystery. Darwin had observed in the *Origin of Species,* "Everyone has heard that when an American forest is cut down, a very different vegetation springs up." One theory held that the appearance of new forests was the product of spontaneous generation (the scientific term at the time was *abiogenesis*): plants and animals simply came into being, wondrously if inexplicably, animated by some mysterious spark that was either chemical or divine. This idea accorded well with the idealistic science of Agassiz, and in 1859 one of his assistants, Henry James Clark, announced that he had observed microscopic animals come into existence from decomposing muscle. At a meeting of the American Academy, "Professor Agassiz corroborated Mr. Clark's statements most fully, and spoke of the discovery as one of the very greatest interest and importance."

Thoreau considered this nonsense. Spontaneous generation was a form of magical thinking. People *wanted* to believe that plants and animals sprang miraculously into existence; they harbored an innate need to find mystery and the supernatural within everyday life. That sort of thinking was unjustified, however, because it ignored the causal relationships that occurred in nature. Since the mid-1850s, Thoreau had filled his journals with observations about the mechanisms that enabled seeds to disperse. He had carefully observed and described burrs, pollen, and maple wings, hypothesizing how each might travel before germinating. As town surveyor, he had measured dozens of woodlots, paying special attention to the saplings struggling to survive in the deeply shadowed undergrowth. He was confident that plants did not spring from nothing.

Scientific controversy may not have been the only thing that occasioned Thoreau's talk. National politics may have played a role as well. The pine tree had been an emblem on the early flags of Massachusetts, but by 1860 it was more commonly associated with the South. Northerners claimed the oak as a symbol of their region, an emblem of hardy and unyielding character. This iconography would become especially prevalent during the Civil War—one early historian of the conflict noted, "Amid the acclamations of the civilized world, our Northern oak struck down the Southern pine." Eight years earlier, in 1852, Charles Loring Brace's best friend, Frederick Law Olmsted, used the Southern pine to criticize the slave economy in his enormous travelogue, *The Cotton Kingdom*. The South, he wrote, was filled with "'old fields'—a coarse, yellow, sandy soil, bearing scarcely anything but pine trees and broom-sedge." For Olmsted, the history of social injustice was legible in "land that had been in cultivation, used up and 'turned out,' not more than six or eight years before." Disregard for the soil had produced "the nakedness of the impoverished earth," from which spindly yellow pines had sprung. The metaphor was obvious: the same disregard for natural resources could be linked to the South's disregard for human lives.

Yet if the struggle between pine and oak forests contained powerful political overtones in 1860, Thoreau did not overtly speak of them. He began his essay instead with an announcement: "I affirmed . . . confidently years ago [that forests are] regularly planted each year by various quadrupeds and birds." If this answer seems obvious to us today, it wasn't at the time. Few people had actually paid attention to New World forestation; even fewer had given the process the sustained attention Thoreau had. That scientists do so now is a tribute to Thoreau's prescience as well as to the environmentalism he helped inspire. But when he announced at the Agricultural Fair that oak forests were planted by animals, he was saying something entirely new. "On the 24th of September, in '57, as I was paddling down the Assabet . . . I saw a red squirrel," he reported. The creature buried an acorn at the foot of a hemlock, and later Thoreau returned to find a sapling growing on the spot. Similar examples were sprinkled throughout his talk. "In short," he explained, "those who have not attended particularly to this subject, are but little aware to what an extent quadrupeds and birds are employed . . . in collecting, and so disseminating and planting seeds of trees."

All of this is interesting enough. But what makes the essay fascinating today is the way its tone abruptly shifts. Thoreau launches into the old quarrel with himself about the adequacy of scientific explanation. Is it possible that science overlooks the fact that nature is directional and alive—is going somewhere? "Nature can persuade us to do almost anything when she would compass her ends," he announces, hinting that the dispersal of seeds might in fact point to an intelligence coursing behind nature. It is as if Thoreau simultaneously accepts Darwin's theory about the way species come into being while rejecting the limits imposed by that theory—as if he were once again trying to rein in the opposing horses of science and transcendentalism that had divided him for so long. For empirical knowledge is finite, Thoreau suggests. After we have exhausted its limits, we are still left with speculation, supposition, and hypotheses. And those are invariably influenced by belief in some ordering principle. For many people, that principle involves a divinity inherited from four thousand years of tradition. But it also mirrors our own ability to order and organize. "There is a patent office at the seat of government of the universe," Thoreau declared now, "whose managers are as much interested in the dispersion of seeds as anybody at Washington can be, and their operations are infinitely more extensive and regular." The image begs several questions: Who or what are these managers? How are they managing the intricate process of distributing seeds? And is there any purpose or goal to their management beyond mere reproduction?

Thoreau was by no means alone in raising such questions. In an early draft of the *Origin of Species,* Darwin had written that nature was composed of "laws ordained by God to govern the universe." Soon after sending his book to Asa Gray, he wrote, "I am inclined to look at everything as resulting from designed laws, with the details whether good or bad, left to the working out of what we may call chance." (Within a year or so he would abandon the idea of design entirely; it was unnecessary, he realized, for his theory.) In September 1860 Thoreau was close to Darwin's position. He assumed the universe was governed by laws, but he also believed that the products of those laws occurred in a more or less random way. He hovered between design and chance, between idealism and materialism. Which is why the next step in his argument in "The Succession of Forest Trees" is so remarkable—for Thoreau locates mystery and wonder *within* materialism.

His touchstone is the seed: an emblem of renewal and vitality. Standing before the audience at the Agricultural Fair, Thoreau announced that he had found some "long extinct plants" growing in the ruins of a cellar. "Though I do not believe that a plant will spring up where no seed has been," he said, "I have great faith in a seed." Here he was dispelling the myth of spontaneous creation, but he was also arguing on behalf of a new kind of magic, a new source of awe. "Convince me that you have a seed there, and I am prepared to expect wonders. I shall even believe that the millennium is at hand, and the reign of justice is about to commence, when the Patent Office, or Government, begins to distribute, and the people to plant the seeds of these things." His point was that nature's fecund banks of seeds bear witness to a world of wondrous scope and intricacy. Millions upon millions of seeds and spores are produced and scattered, broadcast by the air and by animals in order that a few plants may find their niche and grow. The countless complex interactions necessary to produce a single maple may be the result of nothing intelligent, omniscient, or all-seeing. But something almost as wondrous replaces this intelligence: a natural world that is blindly self-directing, a world that is driven by struggle and contingency, a universe authored not by some abstract Almighty—but by itself. The world, Thoreau suggests, is its own autobiography.

"The Succession of Forest Trees" reflects a conflict between two visions. One brims with divinity, the other is purely mechanistic. One carries with it a rich heritage of religious belief, the other whispers that God is redundant amid the promise of new discoveries and more complete knowledge. Thoreau moves fluidly between the two, shuttling between the divine and the here-and-now, between theism and materialism. And he endows each with the other. In the address's final paragraph, he describes seeds as "perfect alchemists I keep who can transmute substances without end." The word *transmute* is important here: it alludes to Darwin's transmutation theory as well as to the potent magic of alchemy, the ancient art of transforming base metals into gold. Thoreau had recently planted some squash in his garden. "Here you can dig," he informed his audience, "not gold, but the value which gold merely represents; and there is no Signor Blitz about it."

He was speaking of Antonio Blitz, a popular magician of the era famous for his ventriloquism, plate spinning, and his so-called "egg bag," a linen sack from which he produced dozens of eggs out of thin air. In this offhand reference to the Welsh-born magician, Thoreau sums up the ambiguities of Darwin's theories in its first year of publication, capturing both the uncertainties and the longings they created. He argues that another form of mystery and magic is still available, one divested of an intervening providence but nevertheless producing wonder at the deep, irreducible materialism of nature.

As such, "The Succession of Forest Trees" is an early response to a world Darwin had introduced—a place divested of God and yet made wonderful by science, a world of weakened faith and exciting discovery. (Emily Dickinson suggested some of the pain of living in this new reality when she wrote, "Nature is a haunted house"; though God once inhabited the natural world, He has since vacated the premises.) Comparing the wonders of seeds and the cheap magic of Blitz, Thoreau concluded with satire: "Yet farmers' sons will stare by the hour to see a juggler draw ribbons from his throat, though he tells them it is all deception. Surely, men love darkness rather than light."

We all believe in magic, Thoreau suggests. We all need to feel that there is something *more*. But the danger is that this need obscures truth. The world *is* filled with magic, Thoreau asserts, *is* rich with mystery—just not the kind that religious tradition has led people to expect and rely upon. In order to experience these things, one has to relinquish certainty, to abandon old faiths and old patterns of belief. One has to live in the nick of time, between orthodoxy and the unknown, searching for knowledge and insight amid perpetual irresolution.

20

Races of the Old World

The Succession of Forest Trees" appeared in Horace Greeley's *New-York Tribune* in early October. Soon dozens of other newspapers reprinted the lecture. Greeley sent Thoreau copies of the essay, with a note challenging some of its conclusions. He refused to relinquish the theory of spontaneous generation. "Friend Thoreau," he wrote, "in the great Pine forest which covers (or recently covered) much of Maine . . . a long Summer drouth has sometimes been followed by a sweeping fire. . . . Not only is the timber entirely killed and mainly consumed, but the very soil, to a depth varying from six to thirty inches, is utterly burned to ashes, down to the very hard-pan." The next season there appeared a new growth of "White Birch—a tree not before known there." Greeley wanted to know how Thoreau reconciled this fact "with your theory that trees are never generated spontaneously, but always from some nut, or seed, or root, preëxisting in that same locality?"

"Friend Greeley," Thoreau shot back. "This is not so much my theory as observation. Yours is *pure theory*, without a single example to support it. As I have said, I do not intend to discuss the question of spontaneous generation, for the burden of proof lies with those who maintain that theory." Instead he informed Greeley that white birches were in fact abundant in the Maine woods, something he knew from having "had occasion to make a fire out of doors there about a hundred times, in places wide apart." This particular birch species produced a minuscule winged seed so abundant that during winter it tinged the snow with its color. "You may infer how seeds get to your burnt land, and I will leave them to sprout of themselves, without telling what extensive birch forests (of the smaller species) I see springing up every year *from those seeds,* especially where the ground has been burned over or plowed."

Charles Loring Brace almost certainly read this epistolary exchange, which was reprinted in Greeley's New York newspaper and which added to the debate between special creation and developmentalism that had raged since the publication of the *Origin of Species.* Brace was a friend of Greeley's, and occasionally he wrote the editor to ask for help finding employment for his orphans. He still recalled the New Year's dinner party at Franklin Sanborn's when Thoreau had evinced such interest in Darwin's new book.

But if he read the exchange—or the essay that prompted it—he failed to grasp Thoreau's crucial distinction between fact and speculation. Throughout the autumn, as he worked on his manual of ethnology, he increasingly struggled to make Darwin's theory fit his own ideas about race and slavery. He increasingly bent facts to fit his own speculations.

He had declared it shameful that "American science in ethnology has become identical with perverted argument for the oppression of the negro." The politics of slavery had polluted dispassionate inquiry, and the "shadow of our national sin has fallen even on the domain of our science, and obscured its noble features to the world." Brace made these comments in the *Independent,* a weekly abolitionist newspaper based in New York and founded by his cousin, the firebrand minister Henry Ward Beecher. In these same pages he had waged his most successful campaigns on behalf of poor children, had defended John Brown, and would soon declare his support for Abraham Lincoln. But for the moment he wished to address the misuse of science by pro-slavery advocates. When ethnologists such as Nott and Gliddon "write costly and elaborate works on Ethnology, whose main object apparently, is to fasten the fetters of the slave, and when such eminent *savans* as Agassiz contribute their articles to them, we cannot wonder that all whose *interest* it is, repeat their arguments and their mistakes, until they become almost axioms to a certain class of minds."

The purpose of his book on ethnology was to correct all this. It would return the study of human societies to the "solid basis of facts and inductive reasoning," applying Darwinian natural selection to the question of race—for "*race* has become a vital [question] to the American people in one shape or another, and it is hardly strange that so many absurdities are continually uttered about it." Perhaps some of those absurdities would be put to rest when it was shown that all races derived from the same human ancestor.

But Brace kept running into difficulties. He had quickly realized that the theory of natural selection could be used against black people as easily as it could be used on their behalf—that in fact Darwinism could be used to support just about any social or political claim one wanted to make. One "important fallacy," he cautioned, "and one repeated *ad nauseam* by our pro-slavery papers, is that *no two very different races can live together, side by side, without the more powerful destroying the weak.*" That black people were destined for extinction unless governed by whites was an argument as old as American slavery. Now it coincided with Darwin's idea that the struggle for existence "almost invariably will be most severe between the individuals of the same species. . . . In the case of varieties of the same species, the struggle will generally be almost equally severe."

This approach failed to take into account several things, Brace argued. First, there was the enormous quantity of time required for real biological change. Less developed peoples often succumbed to more developed peoples—but only at first. "A barbarous race, in contact with a civilized, is very apt to acquire its appetites for stimulants, without the self-control of the more powerful race to govern them," he conceded. But this "degeneracy" was temporary. "House-dwelling people," after all, were "not suited at once to outdoor[s]." Over time "barbarous races" would adapt to "civilized" societies, rising to the same level of culture as the dominant race.

And this led to Brace's second point. Darwin's theory implied that all peoples shared the same basic humanity and were therefore capable of more or less the same level of development. Proof of this shared humanity could be traced through language.

B race made this argument in *The Races of the Old World*, a sprawling, ramshackle work he modeled on the *Origin of Species*. Like Darwin's work, the book is a compendium of miscellaneous facts gathered from a wide range of fields. Unlike Darwin's work, it is deeply marred by a series of internal contradictions.

Its central thesis is simple enough: "There is nothing . . . to prove the negro radically different from the other families of man or even mentally inferior to them." Human history was sedimented in language, buried in grammatical structures like the bones of dinosaurs or the fossilized imprints of trilobites. *The Races of the Old World* proposed to uncover linkages among races through linguistic analysis, relying especially on recent studies such as Franz Bopp's six-volume *Comparative Linguistics,* which carefully traced and analyzed the connections between what had recently come to be known as Indo-European languages.

But if this approach linked dozens of disparate language groups, it also exposed a troubling fact. Linguistics did nothing to suggest that human history was progressive. Instead it chronicled a series of mass migrations, an endless cycle of conflicts and conquests, dislocations and intermixtures. All too often Brace's catalog of invasions and intermingling signaled the *decline* of a culture. The glorious civilization of ancient Greece, for instance, had devolved into what he described as "one of the saddest spectacles which the earth affords, of the weakening and gradual extinction of the power of a race." And while some cultures declined, others never developed at all. Melanesians, Lapps, Magyars, and Hottentots still relied on rudimentary tools and nomadic lifestyles. Why had they not evolved any further?

Like many monogenists, Brace apparently believed that though all humans sprang from the same source, some races had degraded over time.

The reason for this degradation was little understood but probably had to do with climate, diet, war, or disease. Brace described a Portuguese penal colony in the Fernando Islands where the inhabitants "have become so degenerated that they have abandoned agriculture, and do not even possess a boat—a depth of misery which the lowest South Sea Islander have not reached." He described Ireland's poor as a people for whom "two centuries of degradation and hardship are said to have produced physical effects on a population once vigorous and well-formed, which would liken them to the appearance of some of the lowest African and Oceanican tribes."

But this degradation was not the result of inherent failings within a particular group of people. It was the product of oppression and unjust social institutions. Brace blamed prisons and imperial governments. "The African peoples . . . are cursed with the vices and wrongs of Slavery," he reported, while "free-born negro children in Sierra Leone" bore the mark of their liberty in their very physiques, possessing "more intelligent eyes." Missionaries in South Africa had even reported that when black and white children lived in close contact, by "the third generation the shape of the head of the [black] children begins to change."

Human races were not permanent, in other words. The races of the Old World were "a succession of types, rising by almost imperceptible gradations from the low Congo type to the highest black Nubian type and to the brown Tawarek or Berber." One could trace development from the other direction, too, beginning with "the highest Circassian type" and descending "by a series of slight changes to the brown, so that from physical evidence it [is] impossible to decide where one race terminated and the other commenced."

This was race understood in familiar hierarchical terms. (White people invariably inhabit a higher position in Brace's scheme.) But it was also race understood as constantly changing, one type grading into the next, the boundaries of racial categories transient and blurred. The very spectrum of human skin tones obliterated any clear-cut distinctions of race based purely on color. "Scarcely any marks of a human variety are permanent," Brace admitted to his readers. "They continually shade into one another, or are changed or pass away." Using the *Origin* as his source, Brace concluded that skin color was adaptive, a by-product of climate and conditions. The darker color of Abyssinians, for instance, was "due not merely to elevation, but to diet."

All of which raised fundamental questions: What exactly *was* race? Was it simply a biological classification based on physical characteristics? Was it cultural? Geographical? A remarkable fact about *The Races of the Old World* is that it never defines the central term in its title. At times Brace seems to equate race with skin color and hair type. Elsewhere he links it to ethnic, religious, and national traits. In India, for instance, "color and physical traits are not, in that country, distinctive marks of race." But defining race wasn't the biggest problem Brace confronted in *The Races of the Old World*. Imagining a multiracial America was.

He was careful to couch this difficulty in Darwinian terms. The cohabitation of blacks and whites created an evolutionary problem: "Each parent is adapted to a different and peculiar condition of temperature, soil, and climate. The offspring, if it shares these adaptations equally, must be in so far unadapted to its climate and circumstances." This was nonsense, of course. Tens of thousands of free blacks had managed to thrive in the North. But Brace persisted: "That is, a half-blood mulatto in our Northern States, in so far as he has a negro constitution is unfitted for our climate; in the Southern, he is equally unadapted, from his white blood, to the climate there, and it may be several centuries before he becomes suited to either."

Such notions were not uncommon, even among the most enlightened racial thinkers of the nineteenth century. Early in 1861 the radical abolitionist James Redpath, who had published *Echoes of Harpers Ferry* and *The Public Life of Capt. John Brown,* would call for establishing a colony in Haiti where freed American slaves might emigrate. Redpath believed "the people of the Cotton States east of the Mississippi are, in every essential respect, a different and hostile nation to us." He also thought "the fusion of the human races is the destiny of the future." But at the present moment, "the creation of a great Negro Commonwealth in the Antilles is necessary for the elevation of the African race." The commingling of blacks and whites in present-day America was simply too radical even for Redpath.

Brace shared this notion. He believed that all races originated from a common human stock and a single location—just like the Galápagos finches in the *Origin*. Migrating to various regions around the world, humans had adapted to the geographical conditions into which they settled and had transmitted adaptive characteristics to successive generations. Once they had successfully adapted to their setting, there was no need for further development. As a result, the "negro present[ed] his pure type 4,000 years ago, unchanged," and there was now very little likelihood of "his type changing into that of the white."

It's a statement strangely at odds with itself. Brace firmly believed in the emancipation of slaves, and he was equally convinced that blacks and white did not differ in their mental capacities. But claiming that black Americans were incapable of change allowed him to argue against racial mixture. This was important to him because a nation of mixed and biracial children suggested the fragility of whiteness, the impermanence of pure Anglo-Saxon blood. It suggested the unthinkable: that one day America might not be a white nation at all—a possibility distinctly at odds with his belief that Anglo-Saxon America currently represented the apex of human development.

Brace had always believed that America was God's favored nation, a place where democracy and Christianity had fused to create the world's best hope. That vision did not include the South, with its slave economy and its vast population of blacks. In truth, it did not even include New York, where Brace would spend his entire adult life. For him, America was New England. It was Concord, Massachusetts, and Litchfield, Connecticut. It was a place untroubled by racial diversity or mixture, an imagined nation in which difference scarcely existed. This homogenous place was evoked toward the end of *The Races of the Old World* when Brace explained, "There is every reason to believe that in this country, in the warm districts, the negro and white can live side by side without the former diminishing. . . . It is only in the cold latitudes here, that the negro race dwindles away."

Ironically, in adopting this attitude Brace was aligning himself with Louis Agassiz, who immediately after the Civil War confessed his doubts that African Americans should remain in the United States. Blacks and whites were vastly different from one another, he wrote to an acquaintance. That the two races could live together was inconceivable, "one of the most difficult problems upon the solution of which the welfare of our own race may in measure depend." The recipient of Agassiz's letter was the radical abolitionist Samuel Gridley Howe, a member of John Brown's Secret Six, who had written the scientist to ask his opinion on whether "less than two million blacks & a little more than two million mulattoes" were to be "absorbed, diluted & finally . . . effaced." Howe seems to have agreed with Agassiz's judgment.

If Brace relied on Darwin to provide him with a vision of racial unity, he ultimately shared Howe and Agassiz's anxiety about the sudden infusion of black people into northern society. As a result, he retreated into a religious explanation of Darwinian theory, concluding *The Races of the Old World* by arguing that "the great design of the Creator . . . we reverently believe to be, the development of each human being into 'the perfect man in Christ Jesus,' and the building up of an organic *Kingdom of God*." Longing for an age when "at length a Race shall be born, who shall embody and transmit Divine ideas," Brace believed that one day "an organic 'Kingdom of God' [would] be formed among nations, and so the goal of Humanity be reached."

That "Race" was American, and it was white.

21

A Cold Shudder

In the summer of 1860, Charles Darwin wrote his friend and champion Thomas Henry Huxley to discuss the American campaign currently being waged on behalf of his theory. Restless after completing his book, he had hurled himself into a fresh batch of experiments carried out in his secluded home in the Kentish countryside. He planted orchids and dissected insects. He plundered his wife's embroidery basket for brightly colored wools and silk to tie onto the stakes that supported his various plants. He covered her azaleas with netting and draped the beans with gauze. (He didn't bother to tell anyone why he was doing these things.) In the lazy, still afternoons he pollinated primroses with a small paintbrush or attended to the trays of seedlings that were scattered in his study and the outbuildings throughout his estate. In the afternoons he wandered along the path he called the Sand Walk, a trail that skirted a sun-dappled grove near the garden at Down House. He strolled with hands clasped loosely behind his back, head down, pondering.

Throughout July and August he was drawn to sundews: tiny plants that consumed flies and other insects by wrapping sticky leaves around their victims. How did these slow-moving plants trap flies? he wondered. Why did they eat meat instead of using the sun and soil as nutrients, like other plants? Were they in fact some transitional organism between flora and fauna, some strange hybrid creature that embodied features from both? He fed his growing collection of carnivorous plants with scraps of leaves, paper, bits of feather and wood, moss, milk, egg whites, sugar, raw meat. In one harebrained experiment, he paralyzed the plants with doses of chloroform.

Like a fly in the sticky clutches of a sundew, he felt as though he were being slowly digested—or at least that his book was. By July 1860 the *Origin of Species* had been reviewed by nearly every major British quarterly, defended and denounced by England's top men of science, lampooned by cartoonists, and excoriated in the religious press. Richard Owen, the nation's premier primate anatomist, had anonymously attacked the work in the *Edinburgh Review,* lambasting everything from Darwin's ideas and prose style to the friends he kept. Domestic animals, the *Christian Observer* tartly noted, varied only because God wished them to vary. And the popular press enjoyed titillating comments about ape ancestry, the *British Quarterly* printing an imaginary scene in which a monkey proposed marriage to an honest, goodhearted heroine of the sort one might find in a novel by Thackeray or Eliot.

But Darwin also had his supporters. Joseph Dalton Hooker, Charles Lyell, and Thomas Huxley united to defend the theory of natural selection in Great Britain. Huxley was especially pugnacious and would soon become known as "Darwin's bulldog." He believed that the debate over evolution entailed much more than a particular theory. It was an epic battle in the war between science and religion. Huxley stood on the side of clear thought and rational empiricism; with sardonic glee, he blasted theological claptrap and old-fashioned beliefs. "Who shall number the patient and earnest seekers after truth from the days of Galileo until now," he wrote in the *Westminster Review,* "whose lives have been embittered and their good name blasted by the mistaken zeal of Bibliolaters?" Religion had blighted the lives of earnest truth-seekers, but disinterested science would surely triumph in the end. "Extinguished theologians lie about the cradle of every science as the strangled snakes beside that of Hercules, and history records that whenever science and dogmatism have been fairly opposed, the latter has been forced to retire from the lists, bleeding and crushed, if not annihilated; scorched if not slain."

The debate in England came to a head that July in Oxford, at the annual meeting of the British Association for the Advancement of Science. There the Victorian bishop Samuel Wilberforce agreed to speak against the new theory, ponderously expressing his "disquietude" at the possibility of descending from apes. Huxley whispered to a friend, "The Lord hath delivered him into mine hands," and rose onto the platform to reply, "If I would rather have a miserable ape for a grandfather or a man highly endowed by nature and possessed of great means and influence, and yet who employs those faculties for the mere purpose of introducing ridicule into a grave scientific discussion—I unhesitatingly affirm my preference for the ape."

A similar battle was raging in America. Throughout the blistering summer months in Cambridge, as Gray's essays for the *Atlantic* began to appear, Louis Agassiz continued to sputter and rail at the new theory. His response to Gray's important essay for the *American Journal of Science* had still not been published—it had not in fact been written. He promised once and for all to demolish Darwin's theory of natural selection as soon as he could pull himself away from the completion of his grand museum. Darwin wrote about these developments to Huxley that summer. He reported on the American debate over his theory and wryly observed that Agassiz had his hands full with Asa Gray, whose first article for the *Atlantic* would soon be reaching England. "Gray goes on fighting like a Trojan," Darwin wrote. He could scarcely contain his glee.

The *Atlantic* seldom published the names of its authors in those days, but everyone seems to have known that Gray was the author of the July article defending Darwin. Entitled "Darwin on the Origin of Species," it was an instant success. "'Almost thou persuadest me to be a Darwinite—,'" wrote W. H. Harvey, a Dublin algologist and evolutionary skeptic who read the essay as soon as it arrived in Ireland: "not quite, but thou persuadest me to be a Grayite." Darwin himself wrote in late July to say that the article was "uncommonly pleasantly written, & will tell well on the public. . . . My conclusion is that you have made a mistake in being a Botanist, you ought to have been a Lawyer, & you would have rolled in wealth by perverting the truth, instead of studying the living truths of this world."

That same month Gray received a letter from Charles Eliot Norton, the translator of Dante with whom he had discussed the *Origin of Species* in Jeffries Wyman's office on the day after Christmas. Norton thanked Gray for his review, praising its comprehensiveness and fair-mindedness. He had read the article twice and was much taken by Gray's understated differences with Darwin. For Norton it was especially noteworthy that Gray held out the possibility "that specific creation may have had a part in the existing condition of races, as well as variation and natural selection." Norton had immediately understood that Darwin's theory implied a biological connection between humans and apes. He had taken comfort in Gray's statement in the July essay that more evidence was required to settle this issue. In a rare instance of question begging, Gray announced to his readers: "we must needs believe in the separate and special creation of man, however it may have been with the lower animals and plants." Norton welcomed this passage because it reinforced his own opinion that humans were categorically different from beasts: sentient creatures capable of producing Renaissance masterpieces and intricate allegorical poetry. "I wish that you would give [this idea] a fuller treatment in a succeeding number," he wrote.

Like many other readers, Norton was convinced that natural selection failed to explain the special condition of humankind. The theory also failed to explain the transformation of inanimate matter into living beings—the true origin of life. "It is plain that neither separate acts of origination, nor the action of all the principles of derivation taken together will alone explain the facts of organic existence," he continued in his letter to Gray. ". . . It does not seem probable that Darwin's theory, which at first sight appears to do away with specific creation, may lead to an understanding of the laws of origination as well as of those of development." Natural selection might explain the emergence of one species from another, he conceded, but it remained silent on that mysterious moment when matter first flickered into life.

Gray had not addressed the issue of life's origins in his first essay for the *Atlantic*. Norton encouraged him to do so. "Why should not we have the theory of Gray as well as of Darwin," he wrote, "—a theory embracing both specific acts of creation and general principles of derivation, and showing the harmony between them?" Gray agreed to tackle the problem in his second article, which was published in August and titled, like the first, "Darwin on the Origin of Species." There he promised to "inquire after the motives" that impelled Darwin to "press his theory to . . . extreme conclusions" and to propose that *all* living things could be traced to a single primordial organism.

"Why," he asked, "should a theory which may plausibly enough account for the *diversification* of the species of each special type or genus be expanded into a general system for the *origination* . . . of all species?" Part of the answer was simple: Darwin's theory accorded with "great classes of facts otherwise insulated and enigmatic." It explained phenomena that had been previously inexplicable. Why did bird and mammal embryos have gills? Why did two remarkably distinct creatures such as the giraffe and the elephant have the same number of vertebrae in their neck? Community of descent, Gray wrote, explained with sublime eloquence these strange and wonderful occurrences.

Even plants and animals were more similar than commonly assumed. Algae, for instance, were at first "characteristically animal, and then . . . unequivocally vegetable," Gray explained. (He was referring to the ability of algae to consume other plants like herbivores.) Members of the two kingdoms shared other structural similarities, such as the organs of reproduction; conversely, "lower grades of animals" such as amoebas sometimes produced "offshoots" that eventually separated from the parent stock. These characteristics might be taken for homologies, or similarities that were due to relatedness.

Nothing in nature was as clear-cut or as orderly as it seemed. Some plants were bisexual, Gray wrote, others unisexual. The human fetus had characteristics of a fish. The natural world was fluid and uncertain, its categories radically unstable. Gray was not arguing that differences among species, genera, and families were illusory or fictitious. He was pointing out, rather, that any critique of Darwin's theory had to take into account all of the anomalous facts that the theory explained.

But for the first time in his discussion, Gray hesitated. He wavered in his support for Darwin's theory. Natural selection could explain much, he granted, but could it explain *all* of nature's marvels? Was one to believe that chance governed every aspect of the universe? Wasn't it just as likely that some thread of design stitched together pieces of the universe? How else, he asked, explain the eye?

Since at least the time of Socrates, natural philosophers had considered the eye proof that creation had been fashioned by a supernatural being. In 1802 the natural theologian William Paley extended the premise: not only was the eye a remarkably designed optical structure, Paley wrote, but "there is to be seen, in every thing belonging to it and about it, *an extraordinary degree of care, an anxiety for its preservation, due, if we may so speak, to its value and its tenderness.*" For Paley, the eyeball's position beneath thick supraorbital bones, as well as the protection afforded by the eyelid, suggested to him that the Creator considered the eye especially worth protecting.

Darwin had anticipated this argument in the *Origin*. It seemed absurd to "suppose that the eye, with all its inimitable contrivances for adjusting the focus to different distances, for admitting different amounts of light, and for the correction of spherical and chromatic aberration, could have been formed by natural selection." Yet, he continued,

> reason tells me, that if numerous gradations from a perfect and complex eye to one very imperfect and simple, each grade being useful to its possessor, can be shown to exist; if further, the eye does vary ever so slightly, and the variations be inherited, which is certainly the case; and

if any variation or modification in the organ ever be useful to an animal under changing conditions of life, then the difficulty of believing that a perfect and complex eye could be formed by natural selection, though insuperable by our imagination, can hardly be considered real.

The sentence is as carefully constructed as a legal brief. Darwin undermines the canonical argument for design in nature, asserting that random variation explained the formation of the eye just as persuasively as divine fiat. All one had to do was imagine gradual adaptive change occurring over vast stretches of time, from primitive eye-buds to the acute optical organ of a hawk.

Gray struggled with this conclusion. In his copy of the *Origin,* he responded to Darwin's initial statement that such a hypothesis seemed "absurd in the highest degree" by penciling in the margins, "So it does." Earlier that year he had written Darwin to say that "what seems to me the weakest point in the book is the attempt to account for the formation of organs,—the making of eyes, &c by natural selection. Some of this reads quite Lamarckian." He meant that Darwin sometimes implied that animals *willed* the improvement of their vision, much as giraffes, in Lamarck's example, had willed longer necks by stretching them. Descent through modification relied on *accidental* variations.

Darwin replied, "About weak points I agree. The eye to this day gives me a cold shudder, but when I think of the fine known gradations, my reason tells me I ought to conquer the cold shudder." Creatures still existed with rudimentary eye buds, he continued. One could trace the powerful eyesight of predators to these primitive photosensitive nerves. "I feel pretty sure from my own experience," Darwin continued, "that if you . . . keep the subject of Origin of Species before your mind, that you will go further & further in your belief.—It took me long years & I assure you I am astonished at the impression my Book has made on many minds."

But Gray remained skeptical. In his second *Atlantic* essay, he asked how it was possible that blundering, haphazard chance could account for "the most perfect of optical instruments"? How could accidental variation produce this marvel of exquisite form and function? Surely some deeper purpose or design guided its development. "A friend of ours," he noted, privately nodding to Darwin, "who accepts the new doctrine, confesses that for a long while a cold chill came over him whenever he thought of the eye." But Gray was really speaking about himself, not his friend in England.

Something had happened to Gray while he worked on his essays for the *Atlantic*. A wave of unbelief—a surge of dizzying, sickening doubt—swamped him. Darwin's description of a world characterized by endless struggle and prolific death had collided with his religious faith. It called into question God's benevolent hand. Suddenly he felt insignificant: unsheltered and oppressed. Darwin's book rendered nature little more than an empty machine. Sometime that summer Gray's world suddenly became quaint and pitiable.

We know this happened because of the abrupt and strong response it provoked in Darwin. Replying to a letter from Gray that is now lost, Darwin wrote, "In truth I am myself quite conscious that my mind is in a simple muddle about 'designed laws' & 'undesigned consequences.'" Gray had apparently asked whether natural selection might be the mechanism of intelligent design. Might not God use evolution to produce His creation? Darwin confessed his own confusion about the matter. While he resisted embracing a purely material philosophy, he wrote to Gray, he could find little consolation in the traditional Christian explanation of events. The natural world was simply too murderous and too cruel to have been created by a just and merciful God. "I see a bird which I want for food, take my gun & kill it, I do this *designedly.*—An innocent & good man stands under a tree & is killed by flash of lightning. Do you believe (& I really shd like to hear) that God *designedly* killed this man? Many or most persons do believe this; I can't & don't.—"

It was, to put it mildly, a sensitive topic. Gray was a devout Presbyterian, a believer in the Nicene Creed. Throughout his long and illustrious scientific career, he had managed to square his faith in God with his study of the natural world. Sometimes this required him to compartmentalize;

after all, by its very definition, science could never settle questions of the spirit, committed as it was to the study of matter. But until now these two very different modes of understanding the world had always seemed to complement each other. God had created a wondrous planet; the job of science was to study His work and, in so doing, to celebrate it.

Darwin injected doubt into this process. If natural selection produced new species by purely mechanistic means, how did we know the rest of the universe hadn't been created by physical processes? About that man struck by lightning—Darwin asked Gray, "Do you believe that when a swallow snaps up a gnat that God designed that particular swallow shd snap up that particular gnat at that particular instant?" Darwin was perhaps alluding to Christ's words in the Gospel of Matthew: "Are not two swallows sold for a farthing? and one of them shall not fall on the ground without your Father." But he had a different explanation than the Bible: "I believe that the man & the gnat are in the same predicament.—If the death of neither man or gnat are designed, I see no good reason to believe that their *first* birth or production shd be necessarily designed."

Worried he might have offended his American friend, Darwin hastily concluded, "Yet, as I said before, I cannot persuade myself that electricity acts, that the tree grows, that man aspires to loftiest conceptions all from blind, brute force."

Darwin told Thomas Huxley that Gray resembled a Trojan, fighting obsolete science in the name of evolutionary theory. Increasingly, however, Gray began to feel less like a Trojan and more like a Trojan Horse. With his calm and genial voice he had smuggled Darwin's ideas behind the ramparts of American culture—into academies and churches and publishing houses—where they could colonize the minds of America's intellectual class. He had done so willingly and without stint because Darwin's theory was a classic example of evidence-based *science*. But he had also done so because he wanted to place American botany at the forefront of international science and because of his long-standing antagonism toward Louis Agassiz and the racist demagoguery of the polygenists.

Taken together, these factors pushed him to accept Darwin's premises more enthusiastically than he might otherwise have done. They encouraged him to become the foremost American advocate for the theory of natural selection.

What he didn't realize—at least until it was too late—was that the *Origin* was a kind of Trojan Horse, too. It had entered American culture using the newly prestigious language of science, only to attack, once inside, the nation's cherished beliefs. Once the *Origin of Species* gained admission inside a reader's head, it began to compete with all sorts of dearly held convictions, struggling against biblical accounts of the Earth's age and beginnings, destroying consolatory views that had served as a bulwark for much of history. With special and desolating force, it combated the idea that God had placed humans at the peak of creation.

Now, in the summer of 1860, a similar process was happening to Gray. Just as his essays were appearing in the *Atlantic*, he experienced second thoughts about evolutionary theory. As if sensing this change, Darwin wrote in September to say that he hoped Gray would continue to support his theory. "I am thinking of taking a very great liberty; but after much consideration I do not think you can object; you said that *it was known that you were the author of the 1st article;* & as the best chance of getting it reprinted in England in a scientific journal wd be to affix your name, I think of doing this & I hope to Heaven that you will not think this an unwarrantable liberty. I think most highly of this Article & I cannot bear to think it shd not be known in England."

Gray gave his blessing to the venture, but his third essay, published in October, revealed his increasing difficulty aligning Darwin's theory with his own religious convictions. Entitled "Darwin and His Reviewers," Gray's final article began with a concession: "The origin of species, like all origination . . . is beyond our immediate ken." Given the limits of our knowledge, we can only hypothesize about first causes and ultimate beginnings. At present, two explanatory hypotheses were available: "One, that all kinds [of species] originated supernaturally . . . the other, that the present kinds appeared in some sort of genealogical connection with other and earlier kinds." Gray cautioned against settling the point too quickly—"a wise man's mind rests long in a state neither of belief nor of unbelief"—and he elaborated on this point: "Most people, and some philosophers, refuse to hold questions in abeyance, however incompetent they may be to decide them." Something in human nature required answers for even the most difficult questions. "Sometimes, and evidently in the present case, this impatience grows out of a fear that a new hypothesis may endanger cherished and important beliefs. Impatience under such circumstances is not unnatural, though perhaps needless, and, if so, unwise."

Gray now proposed that natural selection *might* be the process by which God had fashioned the world: "Agreeing that plants and animals were produced by Omnipotent fiat does not exclude the idea of natural order and what we call secondary causes," he wrote. With this statement Gray became the first to make a theological case for Darwinian theory. Natural selection, he suggested, might be God's chosen method of creation. This idea would grow increasingly popular in the future because it seemed to resolve the tension between scientific and religious accounts of origins. But it represents a stunning shift for Gray. Before now, he had always insisted that secondary causes were the only items science was qualified to address. First, or final, causes—the beginning of life, the creation of the universe—were the purview of religion: matters of faith and metaphysics.

"Darwin's particular hypothesis, if we understand it, would leave the doctrines of final causes, utility, and special design, just where they were before," Gray told his readers. In the ancient primordial past, inanimate matter had inexplicably become animate. Darwin assumed there was a material explanation for this event; Gray believed that since science could not account for it, there was a chance God had intervened.

He returned, one more time, to the eye. Darwin's theory asserted that all adaptations were "fortuitous or blind." Gray could not accept this. How could "blind forces" produce organs precisely adapted to specific ends? How could blundering accident produce eyes that were "better adjusted and more perfect instruments or machines than intellect (that is, human intellect) can contrive and human skill execute"?

In making this argument, Gray failed to acknowledge a crucial point of Darwin's theory. The eye of whales and hawks and humans had not evolved to enjoy a cerulean summer sky or to glory in God's creation. They had not been fashioned to enjoy the pigments of paint or the delicate tints of autumn leaves. They had developed over aeons in order to plunder, pillage, and overpower: to enable their particular species to survive in the competitive struggle of life. Gray could concede this argument to animals. But he had more difficulty when it came to people. After all, humans were endowed with a moral sense and a perception of divinity.

Gray told his *Atlantic* readers that Darwin's theory, correctly understood, "concerns the *order* and not the *cause,* the *how* and not the *why* of phenomena." It left questions of design untouched. To illustrate this point he employed an image that would soon be used by others in subsequent debates about natural selection. He asked his readers to imagine streams flowing down a slope. The streams were the counterpart of natural selection. They "may have worn their actual channels as they flowed, yet their particular courses may have been assigned; and where we see them forming definite and useful lines of irrigation, after a manner unaccountable on the laws of gravitation and dynamics, we should believe that the distribution was designed."

The passage bears a striking resemblance to Thoreau's description of the "sand foliage" in *Walden.* It shares Thoreau's desire to find a deep underlying law in the seemingly random occurrences of water and soil. But in Gray's analogy the unpredictable movement of streams down a hill is meant to represent variation among species. If streams and variations follow general laws, their mechanisms are unknown. More important, while both resist a strict determinism, both also prove beneficent—Gray's streams produce "definite and useful lines of irrigation." Variations enable species to survive, just as the trickle of streams irrigates the soil. In these obscure operations, Gray professed to see design.

The problem with the analogy was that it used a purely physical process to infer the actions of a Creator. Gray had leaped beyond his own rules of science, speculating about something that was untestable. Abandoning scientific rationalism, he found an intelligent cause "forming definite and useful lines of irrigation, after a manner unaccountable on the laws of gravitation and dynamics."

At some level he must have known that this argument failed to adhere to his own definition of science. But the simple truth was that he found it impossible to live in the world Darwin had imagined: a world of chance, a world that did not require a God to operate. In this way, he was closer to Louis Agassiz than he cared to admit. Like Agassiz, Gray refused to believe that life was a product of pure chance but thought it instead rather like the eye: its splendor inexplicable without a First Cause. "Chance carries no probabilities with it," he wrote in the third essay for the *Atlantic,* "can never be developed into a consistent system, but, when applied to the explanation of orderly or beneficial results, heaps up improbabilities at every step beyond all computation." What Gray meant was that while natural selection might be the process that formed the eye, this process *had* to be guided by a wise and overseeing providence. The alternative was too meaningless to consider. "To us, a fortuitous Cosmos is simply inconceivable. The alternative is a designed Cosmos."

Part IV

Transformations

22

At Down House

One day in the summer of 1862 a tall, angular guest peering with blue myopic eyes through thick spectacles appeared at Darwin's house in Kent, introducing himself with diffident self-confidence. He was happy to make the acquaintance of a man he believed he knew better than most people because he knew the way that man thought. Alfred Russel Wallace never considered Darwin anything but the true originator of the idea of natural selection, the leader in a race in which he had been fortunate to place second. As he told a friend, "Mr Darwin has given the world a *new science,* and his name should, in my opinion, stand above that of every philosopher of ancient or modern times." Darwin remained eternally grateful for the Welshman's generosity and returned the praise whenever he could. "What strikes me most about Mr. Wallace," he confided, "is the absence of jealousy towards me: he must have a really good honest & noble disposition."

Both men assiduously followed news from across the Atlantic, where a war of unprecedented scale and lethality raged with no sign of ending. Darwin was especially attuned to the political divisions that had intensified after John Brown's efforts to inflame a slave rebellion. He agreed with Asa Gray's enthusiasm for Abraham Lincoln, elected president of the

United States without a majority vote. Lincoln's ascendance to the country's highest office was a direct result of Brown's attack on Harpers Ferry, which had divided Democrats and left them without a viable candidate. During the summer of 1860, while debates over Darwin's theory reached their apex, America's pro-slavery press had vilified Lincoln, using his ungainly visage in countless cartoons. Lincoln was portrayed as a suitor of black women or as the missing link between blacks and whites. Sometimes he was even a gorilla ("Ape Lincoln"). In one portrait, Lincoln was described as "Our Next Republican Candidate" whose constituent was none other than P. T. Barnum's What-Is-It? Leaning on a rail, Lincoln says, "How fortunate! that this intellectual and noble creature should have been discovered just at this time, to prove to the world the superiority of the Colored over the Anglo Saxon race; he will be a worthy successor to carry out the policy which I shall inaugurate."

After the election the South immediately revolted. Former Democratic president James Buchanan addressed Congress to complain of the "long-continued and intemperate interference of the Northern people with the question of slavery in the Southern States. . . . The different sections of the Union are now arrayed against each other, and the time has arrived, so much dreaded by the Father of his Country, when hostile geographical parties have been formed." Buchanan was referring specifically to a series of rallies held throughout the South in support of secession, but his language made use of the same powerful metaphors of struggle and extinction that had appeared a year earlier in the *Origin of Species*.

Darwin and Wallace paced up and down the Sand Walk, discussing science and politics. They examined Darwin's garden and sat in the dim study he had fashioned for himself on one side of Down House. They dined together. It is likely that they discussed the topic Darwin had scrupulously left out of the *Origin*—the evolutionary relationship between people and animals. Two years later, in 1864, Wallace published "The Origin of Human Races and the Antiquity of Man Deduced from the Theory of 'Natural Selection.'" The paper specifically addressed the evolution of humans from primates, and it showed how similarly Wallace's thought was to Darwin's. But it was also at this point that the two men's thinking began to diverge and to assume new characteristics of their own.

For example, Wallace did not believe that primitive peoples represented a missing link between primates and humans, as Darwin claimed. He had spent too many years living among the native peoples of the Amazon and Malaysia to believe that the local people possessed lesser mental capacities. The Indian tribes of South America practiced social arrangements every bit as complex as those of England. They engaged in rituals and observed custom. They possessed a rich mythos. These people were not inferior, Wallace insisted, just different: the product of different environments, different historical trajectories, different values and beliefs.

In Wallace's account of human evolution, people first walked on two legs. This freed up the hands, enabling hominids to carry out tasks and create tools imagined by the brain, which expanded in tandem with these developments. (This account remains more or less the one upheld by anthropologists today.) Wallace thought the most exciting aspect of increased brain size was that it rendered physical evolution obsolete. Human intelligence allowed people to manipulate their environments by building houses and planting crops, which in turn allowed them to devote time to higher activities, such as mathematics or the writing of sonnets. They could make jokes and weep at representations of themselves in tragic drama; they could ponder the mysteries of the Spirit.

Which is one reason that by the end of the 1860s, Wallace had become a spiritualist. Like Gray, he too found it impossible to accept Darwin's purely materialist explanation of the universe. It failed, in his opinion, to account for at least three miraculous events in history: the creation of life from inorganic matter, the birth of consciousness in higher animals, and the appearance of moral faculties in humans. Something from "the unseen universe of Spirit," as Wallace called it, had surely had a hand in each of these things, and he would spend the next decade or so gathering evidence to prove that this "unseen universe" existed, attending séances, consulting with spirit mediums, and investigating the paranormal. Increasingly he was convinced there was a higher order of existence and that it was only a matter of time before science discovered it. Within a decade after the publication of the *Origin of Species,* he had come to believe that the human form had long ago stabilized. It was the spirit that continued to evolve. One day the world would be peopled "by a single homogenous race, no individual of which will be inferior to the noblest specimens of existing humanity." Humans would at last escape the iron dictates of natural selection, would free themselves from the imperative to survive at all costs.

Darwin reacted to Wallace's public statements on spiritualism as though to an infanticide. "I hope you have not murdered too completely your own & my child," he wrote. When Wallace wrote an article expressing his latest beliefs in spiritualism, Darwin scribbled a single word in its margins: "No!!!"

W allace had arrived at a position not unlike the one Charles Loring Brace described at the conclusion of his book on ethnology, published in 1863. This was the same year Lincoln issued the Emancipation Proclamation, seemingly fulfilling John Brown's vision of racial equality and consummating Brace's once-radical abolitionist hopes. It was also the same year New York erupted in violent draft riots, confirming the fears of many that white resentment toward black citizens would inevitably result in bloodshed.

Despite the racial violence that tore apart his adopted city, Brace continued to believe that Darwin's evolutionary theory foretold a better future. "The idea of the age is slow growth," he wrote his cousin Henry Ward Beecher in 1869, at the height of Reconstruction, "especially of all moral things. We doubt sudden changes, or at all events, we consider them only feeble beginnings of long-working changes. We do not stand before the great masses of the educated classes and exhort them to a sudden conversion." In 1870 he would continue to publish his ideas on evolutionary theory in the *North American Review* and the *Christian Union.* "In attempting to conceive the divine plans of the great architect, we are of course in a region where human faculties reach but little way; yet it seems a possible conception of an infinite Creator, that He should be able to arrange forces on a general plan, whose particular results He should clearly foresee." For Brace, God realized that the physical laws He created inevitably entailed "future failures and half-effects." Nevertheless the halting development of new species suggested "the great object of Progress and Completeness is being steadily worked out."

This was similar to Gray's effort to synthesize science and religion, to align the workings of natural selection with a sovereign God who determines all things. During one of his perennial visits to Cambridge, Brace admitted to a friend, "We (Dr. Gray and I) generally have incessant disputations and talks on Darwinism." The two men did not shy away from the theory's unsettling implications: "If the soul is a growth from animal faculties and instincts, the probability is less for immortality," he summarized. "Or if the whole universe is an evolution under chance and natural selection from a few atoms in a cosmic vapor, the necessity of God is less." But neither man felt the truth of these hypotheses. A deep and abiding faith led Brace to conclude, "Yet to me Darwinism is not inconsistent with Theism."

As he discussed Darwin's theory, his understanding of it changed. He claimed that "there is no drift toward the worse—no tendency to degeneracy and imperfection. The current of all created things, or of all phenomena, is towards higher forms of life. Natural selection is a means of arriving at the best." This statement contradicted the evidence of human degeneracy he had chronicled in *The Races of the Old World,* but it more conveniently squared with the nation's belief in itself as a progressive force in world history—a belief that took on renewed life after the Civil War. Brace now argued that "if the Darwinian theory be true, the law of natural selection applies to all the moral history of mankind, as well as the physical." Races might become extinct, Brace conceded, but evolutionary theory meant that such losses were part of a larger beneficent framework. Beauty and perfection were the flowers produced by the soil of destruction. It moved him to consider that "each little violet . . . which gladdens our eye on a country walk has depended for its existence on a balancing and interworking of innumerable forms of life during 'ages of ages,' and is the result of laws old as creation." Such insights corresponded "to our highest moral intuition of HIM the 'All-controlling.'"

In July 1872 Brace took a well-deserved vacation from the Children's Aid Society and traveled abroad. He stopped in England to visit the man who had transformed his intellectual life a dozen years earlier. "I am at Darwin's with Mrs. Brace for the night," he wrote a friend back home, extolling the English countryside, which seemed to him calm and therapeutic. The well-tended Kentish countryside stood in stark contrast to the American landscape that, seven years after Appomattox, was still creased and furrowed with mass graves and the sunken ravines where lay the Union and Confederate dead. Here in England was no evidence of the struggle for existence that had nearly dissolved the United States.

Darwin was worn with illness. According to Brace, he worked an hour or so each day, excusing himself after dinner for rest. Then he launched into conversation: about the instincts of dogs and the recent discovery of primitive skulls in California. "He gave one of his lighting-up smiles," Brace recalled, "which seemed to come way out from under his shaggy eyebrows. 'Yes,' he said; 'it is very unpleasant of these facts; they won't fit in as they ought to!'"

Inviting Brace into the study where he had written the *Origin of Species,* Darwin delightedly told him of his hate mail, including "a letter from a clergyman, saying that 'he was delighted to see, from a recent photograph, that no man in England was more like the monkey he came from!' and another from an American clergyman . . . beginning with, 'You d—d scoundrel!' and sprinkled with oaths and texts." These letters amused him, Brace reported, "but not a word did he say of his own success and fame." Instead he exuded good humor and childlike curiosity. "I never met a more simple, happy man," continued Brace, "—as merry and keen as Dr. Gray, whom he loves much. Both he and Lyell think Dr. G. the soundest scientific brain in America."

Gray had last been in England four years earlier, in 1868. He and his wife, Jane, stayed in a cottage near Kew, where he could breakfast with his old friend, Joseph Dalton Hooker, now the director of the Royal

Botanical Gardens, and study sedges from around the world. On weekends the couple visited Down House. Gray found Darwin's household bustling with extended family and a stream of scientists and writers—even the occasional politician—all of whom wished to speak with the great man. According to Jane Gray, Darwin was "entirely fascinating" but also visibly suffering from the illness that plagued him for much of his adult life. "He never stayed long with us at a time, but as soon as he had talked much, said he must go & rest." During the afternoons he showed Gray his property. The two men visited Darwin's kitchen garden, together spading the cold earth in preparation for next spring's crop. Darwin also took his American friend inside his heated greenhouse, built in 1863, during the high point of his interest in orchids, which had supplanted his enthusiasm for sundews. Gray was astonished at his friend's simple workbench; a local cooper or wheelwright could have built the primitive piece of furniture. He watched Darwin sit contentedly, attending to trays of seedlings, repotting plants.

He had been at Harvard nearly thirty years. In all that time he had worked tirelessly to elevate American science, to expand the college's herbarium and develop its science curriculum. He had written dozens and dozens of articles and numerous monographs. He continued this work throughout the Civil War and the unimaginable carnage that resulted, as though incessant work were the only way to cope with the terrible destruction. Gray was childless, but many of his closest friends had sons who fought in the Union Army. When he reported these events to Darwin, he used the language of the *Origin of Species* to describe the conflict: the war was "a struggle for existence on our part," he said, hoping it would prove in the long term that "natural selection quickly crushes out weak nations." Gray generally kept his antislavery sentiments to himself, but as the conflict unfolded, he increasingly felt emboldened to argue that "if the rebels & scoundrels persevere, I go for carrying the war so far as to liberate every negro, tho' what we are to do with this population I see not." To Darwin he also communicated his admiration of Lincoln—"Homely, honest, ungainly Lincoln is the representative man of the country"—and expressed his disgust at the Confederacy, asserting, "The weak must go to the wall, because it can't help it. 'Blessed are the *strong,* for they shall inherit the earth.'" Darwin, who worried that "the South, with its accursed Slavery, shd triumph, & spread the evil," followed this news with great interest, repeating Gray's reports to friends in Europe.

By 1868 the two men were veterans of a somewhat more civil war—the battle over the acceptance of Darwin's theory of natural selection. Darwin had indeed published Gray's three essays from the *Atlantic* in a pamphlet entitled "A Free Examination of Darwin's Treatise on the Origin of Species, and of its American Reviewers." For the pamphlet's subtitle he chose: "Natural Selection not inconsistent with Natural Theology." Appearing in 1861, the work invoked a fresh wave of commentary on Darwin's controversial book. The prominent English theologian Frederic D. Maurice soon observed that "by far the best step forward in Natural Theology has been made by an American Dr. Asa Gray, who has said better than I can all that I want to say." The Scottish botanist J. H. Balfour, a vigorous anti-Darwinian, was gratified that Gray expressed only "a qualified adhesion to Darwin's views."

Through voluminous correspondence the two men continued their private discussion about chance and design. Toward the end of 1860, Darwin wrote, "I had no intention to write atheistically," but he admitted that he could not accept Gray's arguments on behalf of intelligent creation. "I grieve to say that I cannot honestly go as far as you do about design. . . . I cannot think that the world, as we see it, is the result of chance; & yet I cannot look at each separate thing as the result of design." He realized this position placed him at odds with Gray's religious convictions: "I own I cannot see, as plainly as others do, & as I shd. wish to do, evidence of design & beneficence on all sides of us. There seems too much misery in the world. I cannot persuade myself that a beneficent & omnipotent God would have designedly created the Ichneumonidae [a parasitic wasp] with the express intention of their feeding within the living bodies of caterpillars, or that a cat should play with mice." On the other hand, Darwin confessed that he could not "be contented to view this wonderful universe & especially the nature of man, & to conclude that everything is the result of brute force. I am inclined to look at everything as resulting from designed laws, with the details, whether good or bad, left to the working out of what we may call chance."

A similar ambivalence coursed through most discussions of Darwinian theory during the second half of the nineteenth century, in America especially. Religious thinkers like Henry Ward Beecher sought to accommodate natural selection to their progressive theology. Beecher believed that human history was as susceptible to evolutionary forces as nature was. Darwin's theory not only magnified the wonder of Creation, it also seemed to confirm the heavenward ascent of humanity. Social scientists and industrialists, on the other hand, pointed to Darwinian struggle to justify an increasingly stratified American society. Few Americans subscribed to the brutal social Darwinism of Yale professor William Graham Sumner, who observed that "a drunkard in the gutter is just where he ought to be, according to the fitness and tendency of things. Nature has set upon him the process of decline and dissolution by which she removes things which have survived their usefulness." But many people translated Darwin's theory to social issues, some arguing that privileged classes had no obligation to help those who had failed to adapt to an increasingly industrialized modern world. Such arguments not only justified the business interests of the Gilded Age, they exercised especially pernicious effects on race relations during the Reconstruction period and beyond.

Frederick Douglass, white-haired and a little stooped, continued to speak out against America's systemic racism for the remainder of the nineteenth century. He was particularly incensed by the rising incidence of lynching and other violence that began in the 1870s and became a feature of postwar America. Adopting Darwinian language, Douglass described African American existence as a "race of life," a brutal conflict between blacks and whites that had not ended with the freeing of millions of slaves. In his 1894 address entitled "Why Is the Negro Lynched?" he suggested that the outmoded theories of ethnologists had not so much disappeared as returned in new guises. Emancipation had not been enough to guarantee equality for all, because "the spirit of slavery [continued] to perpetuate itself, if not in one form, then in another." Douglass believed that social Darwinism was another pseudoscience designed to oppress free blacks, especially in the South. Partly for this reason, he refrained from endorsing the Darwinian hypothesis that humans and animals were biologically linked. To do so was to invite the same racial stereotypes that had been a feature of American life since its founding.

Through it all, Asa Gray viewed his role in the receding debate over natural selection with mixed feelings. To Darwin he confided that his landmark reviews for the *Atlantic* did "not exhibit anything like the full force of the impression the book . . . made on me"; he had deliberately sought to "stand uncommitted," thinking this a better strategy than "announc[ing] myself a convert." His strategy all along, as he told Joseph Dalton Hooker, was "that Darwin . . . should have a *fair* hearing here."

But this approach had collided with his growing unease about the theological implications of the *Origin*. After the war, as American natural historians and intellectuals increasingly accepted portions of Darwin's book, blending it with elements of Lamarckism, Gray labored to bridge for himself the widening gap he felt between science and religion.

In 1868 Darwin published his *Variation of Plants and Animals Under Domestication*, an enormous, painstaking work that detailed the means by which breeders manipulated the characteristics of vegetables, flowers, and animals. The book's conclusion addressed the very question that had plagued Gray for nearly a decade. As Darwin told Hooker, "It is foolish to touch such subjects, but there have been so many allusions to what I think about the part which God has played in the formation of organic beings, that I thought it shabby to evade the question." Echoing the opening of the *Origin of Species*, Darwin stated that domestic animal breeders and natural selection worked in similar ways. Both favored useful varieties. Both relied on varieties that were *accidentally* produced. He asked his readers to imagine an architect who built a mansion out of rocks that had broken off and splintered at the foot of a cliff. "Can it be reasonably maintained that the Creator intentionally ordered . . . that certain fragments should assume certain shapes so that the builder might erect his edifice?" Like the architect, natural selection constructed something new from fragmentary, accidental materials. Darwin concluded the passage with a reference to his old friend: "However much we may wish it, we can hardly follow Professor Asa Gray in his belief that 'variation has been along certain beneficial lines,' like a 'stream along definite and useful lines of irrigation.'"

Gray admitted that he "was put on the defence by your reference to an old hazardous remark of mine." Darwin's stone house argument could not be answered; "the notion of design must after all rest mostly on faith," he maintained, and faith was something a person felt regardless of evidence or arguments to the contrary. If the two friends discussed the matter during Gray's visit, there is no record of it. Jane Loring Gray, who kept a diary account of the trip, had been particularly impressed with Darwin's face, which bore deeply graven "marks of suffering and disease." He was "tall & thin, though broad framed," and he seemed immensely older than her husband, who was not even two years younger than Darwin.

Eight years later, in 1876, Gray published a collection of everything he had ever written about Darwin's theory. Natural selection was by this time almost universally accepted among American scientists. It was also accepted by a large swath of the liberal clergy, who embraced Gray's arguments throughout *Darwiniana* that natural selection was a mechanism employed by God. Gray no longer insisted upon the separation between science and religion. The origins of the material universe might be beyond our ken, he wrote, but "there are also mysteries proper to be inquired into and reasoned about." Among the mysteries: "Whence this rich endowment of matter? Whence comes that of which we all see and know is the outcome?" These were theological questions—metaphysical questions that resembled transcendentalist speculation more than contemporaneous science. They were wrapped in mystery. Still, Gray noted, most "scientific men have thought themselves intellectually authorized to have an opinion about [them]."

He must have realized that his opinions were beginning to sound old-fashioned and quaint. *Darwiniana* concludes with Aristotle's observation that "the *Divine* it is which holds together all Nature." This idea, he wrote, had "continued through succeeding ages, and illuminated by the Light which has come into the world—may still express the worthiest thoughts of the modern scientific investigator and reasoner." Around this time, and not coincidentally, he was finally invited to join the Saturday Club.

Gray would live another decade, his final year including yet another trip across the Atlantic, where in addition to being conferred honorary degrees by Cambridge and Oxford, he returned once more to study at the Jardin des Plantes in Paris. In the spring of 1882, after visiting the botanist Alphonse de Candolle in Geneva, he learned of Darwin's death. The news staggered him, arriving soon after he learned that both Emerson and Longfellow had also died. About Darwin he could only marvel at his relationship with the man. "We hardly should have thought," he recalled, "twenty-five years ago, that he would have made such an impression upon the great world, as well as on the scientific world!"

Twenty-nine years and four days after the publication of the *Origin of Species* in England, Gray was coming down the stairs of his house on Garden Street when he suffered a stroke. He managed to eat breakfast and then to send off a copy of a review of *The Life and Letters of Charles Darwin,* written and edited by the naturalist's son. But paralysis soon struck, and nearly two months later, in January 1888, he died. At his funeral, Harvard's preacher, Francis Greenwood Peabody, read a scriptural passage that provided a religious gloss to Darwin's notion of the Tree of Life: *Every tree is known by its own fruit: a good man out of the good treasure of his heart bringeth forth that which is good.* To Gray's widow, Charles Loring Brace wrote, "I feel I owe a great deal intellectually to the dear Doctor." Three months later he was still grieving. He considered himself "so much indebted to him for innumerable acts of kindness and consideration, and above all, for the light he threw on so many scientific questions for me and others." He closed with an image of Gray in heaven. "I imagine him living in the highest light of God, and ever learning of His universe."

23

The Ghost of John Brown

For the next four years Louisa May Alcott tinkered off and on with the manuscript of her novel *Moods,* always feeling as though she had failed to achieve the vision of love and passion that had descended upon her during the summer of 1860. In the fall of that year, she emerged from her second-floor room with an ink-spotted draft, which she read aloud to her family. Bronson and Abba were proud. "Father said: 'Emerson must see this. Where did you get your metaphysics?'" She returned to the work in 1862, sitting for hours at her desk or writing on the sofa, where she kept a pillow by her side. If the pillow stood on its end, she was not to be disturbed; if it lay flat, she was taking a break from the tangled affairs of Sylvia Yule and Adam Warwick. Revising the manuscript, she stayed up till "dusk, could not stop, and for three days was so full of it I could not stop to get up." Bronson bestowed his "reddest apples and hardest cider" as reward for such effort. Then in 1864 she once more "worked on it as busily as if mind and body had nothing to do with one another. . . . The fit was on strong & for a fortnight I hardly ate or slept or stirred but wrote, wrote like a thinking machine in full operation." By then she was an established writer, almost a celebrity.

War had made her famous. When fighting between Union and

Confederacy erupted in April 1861, the entire Alcott family believed it was only a matter of time before slavery would be abolished. Bronson noted the firing on Fort Sumter in red ink in his journal, confident the fighting would at last "give us a nationality." By this he meant that the North would at last be united in its devotion to freedom and would no longer compromise its destiny. Concord quickly filled with young recruits in blue woolen coats who drilled in the square and were shipped south to fight the secessionists. Louisa and her mother walked daily to the Concord Town Hall, where a society of women sewed "patriotic blue shirts" for the soldiers. "I long to be a man," Louisa confessed to her diary, "but as I can't fight, I will content myself with working for those who can."

Overnight the country had changed. City streets were filled with soldiers. Red, white, and blue bunting hung from balconies and shop windows. Touring the Charlestown Navy Yard near Boston, Emerson remarked, "Sometimes gunpowder smells good." Ellen, his eldest daughter, wrote, "Here in Concord we all agree that we have had the pleasantest summer that has been known in years." The firing on Fort Sumter had galvanized the town, and she could not recall "a more intimate and social town-feeling, caused by the war probably."

One result of the conflict affected Louisa directly: the *Atlantic Monthly* was no longer interested in her stories. James T. Fields, the magazine's new editor, was keenly attuned to public sentiment; he "has to choose war stories if he can," Louisa explained to a friend, "to suit the times. I will write 'great guns' Hail Columbia & Concord fight, if he'll only take it."

But in order to write about the war, she needed experience. In the winter of 1862 she volunteered to work as a nurse at the Union Hotel Hospital in Washington, a former tavern hastily converted to manage the torrent of wounded soldiers brought in from nearby battlefields. Stationed in an old ballroom, Louisa dressed the wounds of men "so riddled with shot and shell, so torn and shattered," as to have "borne suffering for which we have no name." She would have spent the rest of the war there had she not contracted typhoid fever, an epidemic that eventually killed more than fifty thousand troops in the North and South. A telegraph about her dangerous condition reached Bronson just as he was embarking on a speaking engagement. He canceled the trip and hurried to Washington, where he found his daughter in a narrow iron bed, skeletal and delirious, a woman he barely recognized. "Come home," he said.

During her brief stint in Washington, Louisa had written long, chatty, and surprisingly lighthearted letters home that described her experiences as a nurse. Something about the men she tended—their stoic acceptance of loss, their willingness to sacrifice—lifted her spirits. Bronson forwarded his daughter's letters to Franklin Sanborn, who by 1863 had quit teaching school in Concord to edit an abolitionist newspaper in Boston. As soon as Louisa's letters appeared in print, they attracted a following, quickly becoming so popular that they were published in book form. *Hospital Sketches* proved to be the literary success Louisa had long dreamed of. It even inspired Walt Whitman, also working in Washington's hospitals, to try his hand at an account of the war. In the wake of her newfound celebrity, Louisa set about revising *Moods,* writing at the small semicircular table her father had recently built for her, astonished once again by this feverish product of her imagination. "Daresay nothing will ever come of it," she wrote in her journal, "but it *had* to be done, and I'm richer for a new experience." Hard upon the success of *Hospital Sketches,* her first novel was published in 1865.

It was the *next* novel for which she is best known. Louisa had not particularly wanted to write a book about and for young people. She did so at the suggestion of her publisher. The result was a sentimental account of her own family during wartime—with the notable absence of Bronson. *Little Women* burst upon the postwar literary scene much as *Uncle Tom's Cabin* had a decade or so earlier. It portrayed an indelible family of women—Marmee and her four daughters—who struggle against penury and minor hubris in the cozy warmth of their home. The novel was an instant success, selling out its initial two thousand copies almost overnight. Generations of readers have identified with the March women, who sacrifice Christmas dinner to feed an impoverished immigrant family and gather in sorrowful unity when their beloved Beth dies. Notably absent from the novel is the thematic thread that ran through most of Louisa's works before and during the war: slavery. *Little Women* does not concern itself with the greatest struggle of the era, except indirectly. (Mr. March is a chaplain in the Union Army.) Rather, it addresses the postwar concerns of reconciliation and harmony—concerns from which African Americans were frequently excluded.

As with *Moods,* Louisa was forced to alter the ending of her most popular work. She wanted the novel to conclude with the March sisters as independent women, self-directed and capable of finding fulfillment in work and each other rather than in marriage. Her publisher objected. Louisa complained to a friend that editors could not "let authors finish up as they like but insist . . . on having people married off in a wholesale manner." On New Year's Day 1869, Louisa delivered the second part of the novel, having married off her alter ego Jo to a German professor. If she was disappointed in her compromised ending, she was also suddenly rich and famous, able to pay off the accumulated debts of the Alcott family and to invest the remainder. "My dream," she confided in her journal, "is beginning to come true."

Bronson Alcott would never achieve his daughter's fame, but by the mid-1860s he had achieved a belated prominence that yielded frequent invitations to lecture and attend banquets in his honor. He left Concord to speak in St. Louis and Cincinnati, holding "Conversations" in churches and lyceums throughout the slowly healing nation. The country was changing, its emphases shifting from spiritual and moral concerns to a more pressing and pragmatic materialism. Industrialization accelerated, cities expanded even more rapidly. Amid these transformations, Alcott became a popular speaker on idealism.

Like his daughter, he had been enthusiastic about the war's commencement. The contest between North and South was not a clash just between freedom and slavery but a cosmic struggle between good and evil. Even in 1862, when it had already become clear that the conflict would be protracted and violent, Alcott continued to chant "paeans to the war," which he believed "'was waking the nation to a lofty life unknown before.'" The young writer Rebecca Harding Davis, visiting Concord from war-ravaged Virginia, met Bronson over dinner and concluded, "This would-be Seer . . . knew no more of war as it was than I had done in my cherry-tree when I dreamed of bannered legions of crusaders *debouching* in the misty fields." Not until he traveled to Washington a year later to retrieve Louisa from her sickbed would Bronson confront the true costs of the war. Climbing the stairs to his daughter's attic room, he passed rooms and corridors crowded with amputees and the mortally wounded. Moaning filled the air, and the stench was unbearable. "Horrid war," he admitted to his journal. "And one sees its horrors in hospitals if anywhere."

Yet he remained convinced of the need to fight on behalf of the slaves. In an 1863 lecture about John Brown, Bronson recalled how the old abolitionist impressed him "as a person of surpassing sense, courage, and religious earnestness." Guided by divine intimations, Brown was a superior person, above man-made laws. Alcott considered his actions at Harpers Ferry "the bravest that the country has yet known, the noblest that has been performed for a century; for he risked his life for a race not his own, and struck the first stroke of a revolution that is freeing us all."

Into his eighth and ninth decades, Bronson indulged his penchant for organizing clubs. In 1876 he formed the Fortnightly Club; in 1882 he created the Mystic Club with the express purpose of reading the heretical theologian Jakob Boehme. (Boehme believed, among other things, that the Fall was a necessary stage in the evolution of the universe.) Feeling most at home amid small clusters of like-minded people, he conversed on the same lofty topics that had captured his imagination as a youth, when transcendentalism had consisted of a loose confederation of cranks and dreamers dissatisfied with the everyday commonness of life in America. In those days inspiration and intuition had been of supreme importance—the fleeting recognition of a world beyond this one, a realm shimmering with portent. But as with most movements, transcendentalism had hardened into its own orthodoxy. By 1865 it had attained a tedious respectability. It seemed old-fashioned. After a bloody war waged in the name of ideals, many Americans shifted their focus to the here and now. As the federal government expanded and centralized, as businesses became incorporated, a philosophy that emphasized the solitary individual in nature seemed increasingly irrelevant. Transcendentalism began to feel like an artifact from a previous era, a nostalgic myth.

In 1878 Alcott approached Franklin Sanborn with an idea to combat this trend. He wanted to establish an academy of philosophy, literature, and religion in Concord. The town, after all, was the site of the first battle in the Revolutionary War and the center of abolitionist and transcendentalist thought. What better place to found a school "to which young men and women might resort for the inspiration and insight which our colleges fail to deliver"? A summer course of lectures was planned, and the following year the school opened in the parlor of Orchard House. Two members of the Secret Six were on the faculty (Sanborn lectured on the classics, Thomas Wentworth Higginson on American literature), as well as Benjamin Peirce, the mathematical friend of Louis Agassiz, who had died six years earlier. Emerson was commissioned to lecture on memory, but his own memory was so bad by that point that he was forced to stop halfway through the talk, to sit down and let his daughter Ellen finish for him. Selections from Thoreau's unpublished manuscripts were read aloud.

The school was an instant success. So many students signed up for the first series of courses that a Hillside Chapel was constructed the following year on the slope behind Orchard House. (Louisa contributed much of the endowment to erect the new building.) Modeled on Plato's Academy, the school convened in a large unventilated room still redolent of freshly cut pine, its entrance overseen by marble busts of Plato, Pestalozzi, Emerson, and Alcott himself. Dozens of intellectuals delivered lectures over the next decade or so; thousands of students attended. Alcott's lectures included "Personality," "The Descending Scale of Powers," and "Individualism." The annual reading from Thoreau's unpublished journals quickly became the most popular event of the academy.

If there was an overarching theme to the Concord School of Philosophy, it was that in an era of creeping materialism—of science and technology—idealism was still necessary. Darwinian theory may have helped uncover the physical processes of organic life. It may have even provided support for the idea that all people were linked together. But it took no account of the *soul*. It refused to reference a realm beyond this one, a sphere of transcendent perfection of which this world was but a dim reflection. It thereby diminished the spiritual drama that was part of being human. Alcott sought to combat the effects of Darwinian theory whenever he described the soul's "likeness to the Godhead in his threefold attributes of the beautiful, the true, and the right." He had the *Origin of Species* in mind when he asserted, "Any faith declaring a divorce from the supernatural, and seeking to prop itself upon *Nature* alone, falls short of satisfying the deepest needs of humanity." Like a modern Don Quixote, he determined to combat the tawdry realism of his society by disseminating a philosophy that had shaped his youthful understanding of the world.

If his daughter sometimes found all this high-minded speculation scandalous—"Why discuss the Unknowable till our poor are fed & the wicked saved?" Louisa once wrote of the Concord School of Philosophy—Bronson continued unperturbed, serenely confident that ideals would triumph in the end, that a higher world would one day be revealed. Well into his eightieth year he traveled west to engage in his increasingly famous "Conversations," and after suffering a stroke in 1882, he devoted the last six years of his life to reading and rereading his journals: some five million words chronicling the life of America's most thoroughgoing idealist. Louisa recorded the first word her father spoke as he recovered from the stroke, and that word says everything about Bronson Alcott's enduring perspective during an era of rapid change. The word was "up."

If Alcott's idealism was the presiding impetus for the Concord School of Philosophy, it was Franklin Sanborn's organizational talents that made it a reality. He scheduled courses, handled tickets, oversaw the printing and publication of lectures, and even helped direct the construction of the Hillside Chapel. As a member of the school's faculty, he lectured on Greek oracles and Latin poetry. Typically he was the person who concluded the session by reading from Thoreau's manuscripts.

These were by no means his only activities. After the war he edited the *Springfield Republican* from 1868 to 1872, contributing weekly columns to the paper for years afterward. In 1863 the governor of Massachusetts appointed him secretary of the newly formed State Board of Charities, where he instituted a system of inspections and reporting that would soon be copied throughout the nation. From 1879 to 1888 he was the General State Inspector of Charities. During this time he was also an active officer in dozens of other organizations, including the American Social Science Association, the Clarke School for the Deaf, the National Prison Association, and the National Conference of Charities and Corrections. He lectured widely. He wrote more than twenty books. He collected money to relieve the plight of the urban poor and helped reform the prison system. For thirty decades he edited the *Journal of Social Sciences.*

Three years to the day after Sanborn hosted the dinner party for Charles Loring Brace—January 1, 1863—Abraham Lincoln signed the Emancipation Proclamation, effectively making official what had become increasingly apparent to most observers: that the Civil War would end not in compromise but in the complete social transformation of whichever side lost. At this precise moment Sanborn did what many radical abolitionists did—he turned his attention away from issues of race and slavery. Like most New Englanders, Sanborn knew very few black people. His abolitionist activities were made in the service of an ideal rather than on behalf of specific individuals. When the war ended, he seems to have believed that his work to free the slaves was complete. He afterward concerned himself very little with the struggle of black people to become integrated into white society.

As with so many of his generation, the Civil War marked a permanent division in his life, a gleaming white line over which he could peer but not cross. At the same time, the War of Southern Rebellion would always seem anticlimactic to him, the aftermath to a thrilling battle against slavery he had fought since his mid-twenties. If the Emancipation Proclamation was the formal victory, Sanborn paid scant attention to Reconstruction. He did not weigh in, for example, when in 1865 the term *miscegenation* was coined to provide a scientific respectability to a social arrangement previously known as *amalgamation*. (*Miscere* is Latin for "mixture," *genus* for "people.") The neologism signaled a new stage in the relationship between science and racism, encouraging many states to institute antimiscegenation laws that made intermarriage between blacks and whites a crime. Darwin's evolutionary theory would soon be co-opted by Southerners and Northerners alike to "prove" black inferiority. African Americans and whites might be from the same species, the argument went, but clearly the two races were in an ongoing struggle for existence, and just as clearly whites were better adapted for ultimate survival. Emancipation did nothing to diminish the belief that "the Negro would disappear beneath the glare of civilization," as the abolitionist lawyer Albion Tourgée observed, ". . . a simple and easy solution to a troubling question." Darwinian thinking became yet another justification for laws that restricted black rights.

After the war Sanborn spent less time thinking about the future and instead inhabited the past. Like Bronson Alcott, he devoted himself to genealogy, collecting a small library of histories devoted to Massachusetts and his home state of New Hampshire. The deaths of old friends pushed him deeper into retrospection. When Emerson died in 1882, Sanborn organized a special series at the Concord School of Philosophy to commemorate the great man. Bronson and Louisa May Alcott both died six years later, in 1888, and with them went the Concord School of Philosophy. Walt Whitman, who had sat in a Boston courtroom thirty years earlier and cheered Sanborn after he had resisted arrest, died in 1892. More and more it seemed as if an entire era were hurtling toward its conclusion.

Alone with his memories, Sanborn became the self-appointed caretaker of the past, a solitary groundskeeper in the transcendental graveyard. Increasingly referred to as the "last transcendentalist," he wrote a series of adulatory and often inaccurate reminiscences of the giants of his youth: Emerson, Alcott, Thoreau, Hawthorne, and especially John Brown. He was becoming extinct himself, a living fossil, but he continued to defend

John Brown's epochal insurrection and to proclaim his own role in supporting the raid on Harpers Ferry. A struggle between two conceptions of America—between two antithetical ways of life—had dominated the years before the Civil War. Sanborn had come down on the right side of history, had believed that the struggle for existence was necessary. He had acted when action was paramount.

Fifteen years or so after Brown's execution, he helped collect a small trust for the martyred man's family. Money was gathered from Emerson, Thomas Wentworth Higginson, the poet John Greenleaf Whittier, and many others and sent to Brown's widow, who had moved to Santa Clara County, California. In 1885 he published his *Life and Letters of John Brown,* a biographical account of the abolitionist. "I knew John Brown well," Sanborn explained. "He was what all his speeches, letters, and actions avouch him,—a simple, brave heroic person, incapable of anything selfish or base." More important, "he was an unquestioning believer in God's fore-ordination and the divine guidance of human affairs." Twenty years later he was still pondering the man, who filled a disproportionately large portion of his two-volume autobiography, *Recollections of Seventy Years.* And as late as 1916—a year before his death—he observed, "I knew John Brown for a few years only, but I knew him intimately. From the first I honored him and the more I learned of his life the more I honored him."

He was haunted by the man's self-immolation, his cheerful surrender to the gallows in the name of an ideal. This willingness to die for a cause posed difficult questions for Sanborn. Was it not one's duty to continue fighting? To struggle? To overcome? Sanborn could never entirely shake the idea that to shape the world, one had to first survive it.

Sometime in the late 1870s he made his way to Brown's abandoned house in the Adirondacks of North Elba, New York. Brown's body had been carried back to this home and buried just beyond the spare unpainted house two decades earlier, his grave marker a huge stone with the simple legend JOHN BROWN. 1859. Sanborn slept alone inside the empty house that night, but he was awakened shortly after midnight. The room was as dark as a swamp. In a corner he saw a hallucinatory shape, a body, heavy and inert, swinging from the end of a creaking noose. In the letter he wrote to his son describing the experience, Sanborn confessed his belief that he had seen John Brown's ghost that night.

What he didn't say was that that body might just as easily have been his own had he succumbed to the men who came to arrest him years ago in Concord. He still remembered the surge of terror, the panic he had felt when one of the visitors pulled out a pair of handcuffs. To surrender at that moment would have been to swing from the end of a noose in Virginia. It would have been to forfeit that future into which he had imagined himself striding forever.

That April night in 1860 had changed Sanborn. It made him more cautious, narrowed his boundaries. It also left him exhausted. For a moment, while he struggled in the arms of his captors, he must have felt his selfhood obliterated, his interior life nullified by the capture of his body. He must have felt like a slave. Like a cornered animal, he had thrashed and writhed, shouting for help, kicking to splinters the door of the coach that waited outside his house to whisk him to Washington. He did not want to die—not for an ideal, not for a cause, not for anything. If he learned nothing else from those days of turmoil and agitation, he learned that simple lesson. He would struggle to survive.

24

In the Transcendental Graveyard

That day in early December 1860, Thoreau put on a heavy coat and his broad-brimmed hat and prepared to enter the woods. He walked south to Fair Haven Hill, a sloping promontory that overlooked Walden Pond, and counted the rings of trees that had recently been cut for firewood. Since October he had been engaged in this exercise of dendrochronology, visiting one woodlot after another, trying to determine the age of the trees, how they had been planted, what species had preceded them. His journal had become a biography of forests.

It had been a year of discovery, a year of wonder. Three hundred and sixty-six days earlier John Brown had been executed, an event as freshly astringent in his memory as though it had happened that morning. Perhaps because of this anniversary, Brown was on Thoreau's mind that day. Later in the afternoon he encountered two of his less enlightened townspeople, who bluntly informed him that the old abolitionist had been wrong to throw away his life so casually. No man, they said, "had a right to undertake anything which he knew would cost him his life." Thoreau fumed. He asked "if Christ did not foresee that he would be crucified if he preached such doctrines as he did." The two men, seeing that they had been trapped once again by the irascible hermit, said perhaps not. Thoreau let the matter drop. "Of course," he wrote, "they as good as said that, if Christ *had* foreseen that he would be crucified, he would have 'backed out.'"

Normally Thoreau would have discussed these matters with Bronson Alcott. It was his habit to stop by Orchard House on his way out of town to sample Alcott's excellent cider and inquire about the women of the house. "I think that I love society as much as most," he had once confessed, disputing his reputation as a hermit, "and am ready to fasten myself like a bloodsucker for the time to any full-blooded man that comes in my way." He loved to visit Alcott's study, to warm himself by its fire, his Yankee drawl contending with the mute busts of Plato and Socrates that gazed from their shelves. The two friends discussed gardening, philosophy, and especially this season, politics. On November 6, Alcott noted in his journal, he had ventured to the "Town House and cast my vote for Lincoln, and the Republican conventions . . . the first vote I have cast for a president." Thoreau had done the same, and later that evening the two had cheered the Republican's improbable victory.

But on this day Alcott was confined to his bed, too sick to see visitors. He first complained of a cold on November 29, which was the last time Thoreau had visited. The following week his journal was mute except for a single sentence, less legible than usual, scrawled on December 1: "Confined to my room with influenza." Now, two days later, Thoreau skipped the visit to Orchard House and walked out of town.

The endless debate between science and art continued to rattle inside his head. A month or so earlier he had gazed at the surrounding countryside, exhilarated by a view enriched with his detailed knowledge of plant species. "No one but a botanist is likely to distinguish nicely the different shades of green with which the open surface of the earth is clothed,—not even a landscape-painter if he does not know the species of sedges and grasses which paint it." But he was soon taking the other side in this quarrel, arguing on behalf of the imagination. "The scientific differs from the poetic or lively description somewhat as the photographs, which we so weary of viewing, from paintings and sketches, though this comparison is too favorable to science." Science, he now felt, was a makeshift structure, a quixotic effort to describe and understand the world with limited tools. The truest description of a dandelion was not its weight or its color as measured on a spectrum, but "the unmeasured and eloquent one which the sight of it inspires." Analysis failed to capture the relationship between flower and viewer, the tangled relationship of thought and feeling that occurred when one gazed at a dandelion and experienced a yellow burst of sunlight within. Such ineffable moments eluded scientific description.

Still, Darwin continued to alter his thinking. The *Origin of Species* had inserted itself into his life and permanently reoriented his understanding of the woods and meadows surrounding Concord. One golden October day Thoreau walked to Beck Stow's farm and examined a little pool choked with lilies and pickerelweed (he now used the scientific name *Pontederia* in his journals). Darwin's ideas about variation added a new layer of insight to what he saw: "I have no doubt that peculiarities more or less consider-able have thus been gradually produced in the lilies thus planted in various pools in consequence of their various conditions, though they all came originally from one seed."

This observation led to others. He wondered how these particular plants had first come to this tiny isolated pool. There was no spring or creek to explain it, no pickerelweed nearby. Attacking Louis Agassiz's theory of special creation, which he increasingly thought denied all that was mobile and fluid and enchanting in the world, Thoreau believed that the tiny pool at Stow's farm was no doubt stocked the way "this town was settled," and "we are warranted . . . in supposing that . . . there was not a sudden new creation" in Concord, either.

With this statement Thoreau traveled further than Charles Loring Brace and Bronson Alcott, further even than Asa Gray. To varying degrees, Darwin's earliest readers in America had found themselves incapable of embracing the full implications of the new theory. Sanborn, who gave little thought to the larger philosophical issues, had embraced the theory because it suited his understanding of life in America as a never-ending struggle. But Gray, Brace, and Alcott had each in his own way considered the meaning of a world governed by natural selection. They all tried to accommodate evolutionary theory to their commitments and beliefs, at times finding remarkable convergences. But all three recoiled from those elements of the theory that removed providential design from the universe and demoted human beings to the status of animals.

A century after the publication of Darwin's book, the great science writer and intellectual heir to Thoreau, Loren Eiseley, would write of human evolution, "We are rag dolls made out of many ages and skins, changelings who have slept in wood nests or hissed in the uncouth guise of waddling amphibians. We have played such roles for infinitely longer ages than we have been men. Our identity is a dream." Thoreau was suggesting something similar when he noted in his journal that humans were the product not of miraculous intervention but of lineal descent and geographical distribution. Like dandelion fluff blown hither and yon, populations had scattered around the world, metamorphosing over infinite stretches of time. Darwin's explanation for the appearance of plants and animals might not account for the ineffable experience a person had when falling in love or smelling the first daffodil of the spring, but it *did* account for the way plants and animals and *people* had appeared on Earth and then dispersed themselves over countless eons.

Darwin's book had also prompted Thoreau's latest passion, the meticulous recording of forest trees. On October 19 he noted that he had found the "stumps of the oaks which were cut near the end of the last century." These trees, as best he could tell, had been replaced by three successions of forests. "Perhaps I can recover thus generally the oak woods of the beginning of the last century, if the land has remained woodland." He realized that he possessed an advantage over geologists, because he could determine not only the order of events but, thanks to the tree rings, the precise time during which they happened. "Thus you can unroll the rotten papyrus on which the history of the Concord forest is written."

He was, in other words, adding nuance and complexity to Darwin's theory of nature. It was true that pine and oak forests contended for resources in an endless battle to survive. ("The pines are the light infantry, . . . the scouts and skirmishers; the oaks are the grenadiers, heavy-paced

and strong, that form the solid phalanx.") But it was also true that reciprocal relationships emerged from this struggle. Pine forests protected young
oaks from frost, allowing them to develop into saplings. Competition
enriched both species. Moreover, *human* involvement was often just as crucial as competition to the success or failure of a particular tree. It was local
custom to allow pines to spread in cow pastures. "But when, after some
fifteen or twenty years, the pines have fairly prevailed over us, though they
may have suffered terribly and the ground is strewn with their dead, we
then suddenly turn about, coming to the aid of the pines with a whip, and
drive the cattle out."

Many years later Alfred Russel Wallace would provide "illustrative
cases of the struggle for life" to defend Darwin's theory from skeptics. One
example was "the struggle between trees [birches and beeches] in the forests of Denmark." Thoreau anticipated this remark by a quarter-century.
He carefully recorded the ways in which oaks and pines advanced,
retreated, produced hundreds of thousands of acorns and pinecones to
gain a foothold in a creation already crammed to the bursting point with
life. Winged seeds, barbed seeds, thistles, and milkweed down: each had
adapted to propagate and thrive in an environment that was constantly in
process. With Darwinian natural selection in mind, Thoreau wrote, "The
development theory implies a greater vital force in nature, because it is
more flexible and accommodating, and equivalent to a sort of constant
new creation."

Constant new creation. The phrase represents an epoch in American
thought. For one thing, it no longer relies upon divinity to explain the
natural world. Only a decade earlier Thoreau believed he had found evidence of a divine "Artist" working in the unexpected medium of thawing
mud. He had been encouraged to think this way by Emerson, who claimed
that every physical fact was the outward representation of a spiritual truth.
Emerson had prodded Thoreau to look *through* nature—not at it—in
order to perceive the godhead. To a degree, Thoreau had always resisted
this approach; he loved the hard surface of things too much. But now,
within the short span of a year, Darwin had propelled him toward a

radically different vision of creation that could be explained *without* an Artist. "The development theory" suggested a natural world sufficient unto itself—without the facade of heaven. There was no force or intelligence behind Nature, directing its course in a determined and purposeful manner. Nature just *was.*

Thoreau would continue to develop these ideas into the next year, when the concept of *secession*—not succession—preoccupied most of the nation. "It is a vulgar prejudice that some plants are 'spontaneously generated,'" he wrote in March 1861, one month before the start of the Civil War, "but science knows that they come from seeds, *i.e.* are the result of causes still in operation, however slow and unobserved." If divine intervention was unnecessary to produce plants, then the world was governed by luck and coincidence to a degree most people refused to acknowledge: "Thus we should say that oak forests are produced by a kind of *accident.*"

We will never know how far Thoreau might have absorbed and extended Darwin's theory. Nor can we know what insights he might have extracted from applying the principles of variation and natural selection to his beloved woods and fields. An accident of another sort befell him during the final month of 1860. "I took a severe cold about the 3d of December," he noted in his journal, "which at length resulted in a kind of bronchitis, so that I have been confined to the house ever since."

Most likely he contracted influenza from Alcott—it was a particularly bad winter for the illness, and he had visited when the philosopher was still contagious. Thoreau managed to deliver a lecture in Waterbury on December 15, his voice ragged and wheezy as he spoke. Then he remained in bed until nearly Christmas, impatient to return outside and continue his investigations. While he lay in bed, South Carolina declared that "the union now subsisting between South Carolina and other States, under the name of the 'United States of America,' is hereby dissolved." Other Southern states soon followed suit, and by the spring of 1861 the once impossible nightmare of civil war seemed increasingly likely.

But Thoreau did not recover. The influenza had apparently exacerbated a dormant case of tuberculosis, a scourge of New England that had already carried off one of his sisters. For the next year or so he tried intermittently to venture into the woods, to visit the placid blue-green waters of Walden, to sit outside and absorb the sunlight. More often than not weakness assailed him, and he was forced to turn back. His cough grew worse. Alcott visited his friend and observed that being housebound was "a serious thing to one who has been less a housekeeper than any man in town, has lived out of doors for the best part of his life, has harvested more wind and storm, sun and sky, and has more weather in him than any."

Just before the Battle of Bull Run produced the first Union defeat in the summer of 1861, Thoreau embarked on a long train trip to Milwaukee, hoping the climate would improve his health. It did not. At some point he realized he would never recover, that it was time for a strict closing of accounts. He embarked on the bittersweet process of putting his manuscripts in order, extracting from his sprawling notebooks a few essays for the *Atlantic*. By the following spring he was a wraith. In March 1862 he wrote an admirer of *Walden*, "I *suppose* that I have not many months to live; but, of course, I know nothing about it. I may add that I am enjoying existence as much as ever, and regret nothing." A friend, Sam Staples, visited Thoreau a few days later and reported to Emerson that he had never seen "a man dying with so much pleasure and peace."

The spring of 1862 was refulgent with new growth and renewal. It was also shadowed by the resumption of fighting between Union and Confederate forces after winter. That March Thoreau could barely see or speak. He could no longer hold one of his family's pencils. His cumbersome breathing filled the second floor of his mother's house, where he had been moved. "Elizabeth Hoar is arranging his papers," Abba Alcott wrote her brother, Samuel May, on March 24, 1862, "—Miss [Sophia] Thoreau copying for him—he is too weak to do any of the mechanical part of himself."

At some point his aunt visited his sickroom to ask, "Have you made your peace with God?"—to which Thoreau replied, "We never quarreled." When another acquaintance asked, "Are you ready for the next world?" his response was sardonic and mirthful: "One world at a time."

He died on May 6, 1862. News spread fast throughout Concord, prompting Sarah Alden Ripley to lament, "This fine morning is sad for those of us who sympathize with the friends of Henry Thoreau, the philosopher and the woodman." Bronson Alcott visited during his final day, describing Thoreau as "lying patiently & cheerfully on the bed he would never leave again. He was very weak but suffered nothing & talked in his old pleasant way saying 'it took Nature a long time to do her work but he was most out of the world.'" Against Thoreau's wishes, Emerson arranged a funeral service at the First Church, his sorrow "so great he wanted all the world to mourn with him." Speaking at the memorial, Emerson said of his friend's unfinished natural history manuscript, "The scale on which his studies proceeded was so large as to require longevity, and we were the less prepared for his sudden disappearance. . . . It seems an injury that he should leave in the midst his broken task, which none else can finish,—a kind of indignity to so noble a soul, that it should depart out of Nature before yet he has been really shown to his peers for what he is."

Louisa May Alcott composed a poem for the funeral entitled "Thoreau's Flute" in which those left grieving receive a message blown through the dead man's prized instrument bidding them not to despair. Once when Louisa was a girl, Thoreau had taken her to see a spider's web glittering with dew, telling her it was a fairy's handkerchief. Now, Louisa wrote, Thoreau was not so much dead as returned to the natural world he loved so well: "A potent presence, though unseen,— / Steadfast, sagacious, and serene: / Seek not for him,—he is with thee."

None of this was apparent on December 3, 1860, a day that glowered with thick, plum-colored clouds. Thoreau was living in the nick of time, continuing his arduous pilgrimage to know the world. He moved from tree stump to tree stump, squatting, counting the rings of a pine whose diameter was nearly three feet wide, taking notes in the little booklet he kept in the pocket of his overcoat. He ignored the tickle in his throat.

He would have heard the rain before he felt it, a pattering on the forest roof, a rattling amid the pines and hickories and sumac branches just beyond the field of stumps. Then a drop or two struck the brim of his hat. In *Walden* he described a spring day when he had felt unaccountably lonesome in the quiet solitude of the woods. It had begun to rain, and with this change in atmosphere, his perspective had miraculously transformed. "Every pine needle expanded and swelled with sympathy and befriended me. I was so distinctly made aware of the presence of something kindred to me, even in scenes which we are accustomed to call wild and dreary, and also that the nearest of blood to me and humanest was not a person nor a villager, that I thought no place could ever be strange to me again."

He left off counting tree rings now as his coat became wet. He would return to this spot tomorrow, and the day after that, if necessary, for his self-appointed task. Surely he felt alive at that moment—in a present that was constantly generating its own future, unfolding through time, producing new life even when it seemed dormant and dead. The air smelled of mold and decayed leaves, of the astringent odor of pines. The world, inhabited by an endless variety of plants and animals and people, seemed suspended in fragile balance. It was enough to observe this singular moment of change, a tiny incident in an endless tale of constant new creation of life. This world was enough.

It rained harder as he walked home, and he shuddered in the cold.

Acknowledgments

Many thanks to those who have listened to my ideas and read my manuscript—
Juliet Fuller, Marianne Merola, Paul Slovak, Lars Engle, Bob Jackson, Joli
Jensen, Brian Hosmer, Kristen Oertel, Ken Taylor, Laura Dassow Walls,
Bill Rossi, and John Stauffer—

*and to the generous institutions that made those ideas and that manuscript
possible*—the John Simon Guggenheim Foundation and the National
Endowment for the Humanities—

*and to those libraries and archives that granted permission to quote and
use images*—Harvard's Houghton Library, Gray Herbarium, Ernst Mayer
Library of the Museum of Comparative Zoology, and Countway Library
of Medicine; the Concord Free Public Library; the New York Historical
Society; the spectacular Berg Collection at the New York Public Library;
the Library of Congress; the Morgan Library & Museum; the New Bedford Museum of Whaling; and the National Portrait Gallery, Meserve
Collection.

To all, as Thoreau put it, "my thanksgiving is perpetual."

Notes

ABBREVIATIONS OF ARCHIVES

CFPL: Special Collections, Concord Free Public Library

GH: Gray Herbarium, Harvard University

HL: Amos Bronson Alcott Papers, Houghton Library, Harvard University

MCZ: Archives of the Museum of Comparative Zoology, Ernst Mayr Library, Harvard University

MLM: Morgan Library and Museum

NYHS: The Victor Remer Historical Archives of the Children's Aid Society, New-York Historical Society

NYPL: Berg Collection, Manuscripts and Archives Division, New York Public Library

ABBREVIATIONS OF BOOKS AND PERIODICALS

Correspondence: The Correspondence of Charles Darwin, 21 volumes

Diary: Amos Bronson Alcott diary, HL

HDT: Henry David Thoreau, *A Week on the Concord and Merrimack Rivers / Walden; or Life in the Woods / The Maine Woods / Cape Cod.* New York: Library of America, 1985.

HDT-CE: Henry David Thoreau, *Collected Essays and Poems.* Edited by Elizabeth Hall Witherell. New York: Library of America, 2001.

Journal: Henry David Thoreau, *The Journal, 1837–1846, 1850–November 3, 1861.* Edited by Bradford Torrey and Francis Allen. Boston: Houghton Mifflin, 1906.

LAG: Asa Gray, *Letters of Asa Gray.* Edited by Jane Loring Gray. Boston: Houghton Mifflin, 1894.

LCLB: Charles Loring Brace, *The Life of Charles Loring Brace: Chiefly Told in His Own Letters.* Edited by Emma Brace. New York: Charles Scribner's Sons, 1894.

Origin: Charles Darwin, *On the Origin of Species: A Facsimile of the First Edition.* Cambridge, Mass.: Harvard University Press, 1964.

ABBREVIATIONS OF NAMES

ABA: Amos Bronson Alcott

LMA: Louisa May Alcott

CLB: Charles Loring Brace

CD: Charles Darwin

RWE: Ralph Waldo Emerson

AG: Asa Gray

FBS: Franklin Benjamin Sanborn

HDT: Henry David Thoreau

Chapter 1: The Book from Across the Atlantic

3 **"that of opposition to whatever"** Edward Stanwood, "Memoir of Franklin Sanborn," *Proceedings of the Massachusetts Historical Society* 51 (1918): 310–11.

5 **"what portion of my income"** *HDT,* 325.

5 **"Manual of Ethnology"** FBS to Theodore Parker, January 2, 1860, CFPL.

6 **"John Brown's martyrdom"** Cynthia H. Barton, *Transcendental Wife: The Life of Abigail May Alcott* (Lanham, Md.: University Press of America, 1996), 165.

7 **"did not recognize unjust human laws"** *HDT-CE,* 407.

7 **"it wasn't Brown that was hanged"** FBS to Theodore Parker, January 2, 1860, CFPL.

7 **"the Messiah of the black race"** ABA, *Diary* (1860), 5, HL.

8 **"strengthened the Republicans"** FBS to Theodore Parker, January 2, 1860, CFPL.

8 **"will continue as long"** John Stauffer and Zoe Trodd, *The Tribunal: Responses to John Brown and the Harpers Ferry Raid* (Cambridge, Mass.: Harvard University Press, 2012), 135.

8 "I *never* eat cheese" Benjamin B. Hickok, *The Political and Literary Careers of F. B. Sanborn,* Ph.D. diss., Michigan State University, 1953, 431.

8 "vile outpourings of a lewd woman's mind" Quoted many places including James Penny Boyd, *Triumph and Wonders of the 19th Century: The True Mirror of a Phenomenal Era* (Philadelphia: A. J. Holman, 1899), 123.

8 "'Adam Bede' has taken its place" Rebecca Mead, *My Life in Middlemarch* (New York: Crown, 2014), 15.

9 "savage instinctive hatred" CD, *Origin,* 202–3.

9 "If we admire several ingenious contrivances" Ibid., 203.

10 "just published Darwin's Book" ABA, *Diary,* 1860, 5, HL.

11 "Mr Alcott and Mr Thoreau" FBS to Theodore Parker, January 2, 1860, CFPL.

11 "a theory of moral" *LCLB,* 285.

12 "To read well" *HDT,* 403.

12 "A man receives" HDT, *Journal,* 13: 77.

Chapter 2: Gray's Botany

13 "book is out" A. Hunter Dupree, *Asa Gray: American Botanist, Friend of Darwin* (Baltimore: Johns Hopkins University Press, 1988), 267.

13 "If ever you do read it" CD, *Correspondence,* 7:369.

13 "fragrant as his flowers" James Russell Lowell, "To A.G.: On His Seventy-Fifth Birthday," *Botanical Gazette* 11 (1886): 9.

14 "one spirit reigns" John F. W. Herschel, *A Preliminary Discourse on the Study of Natural Philosophy* (Chicago: University of Chicago Press, 1987), 4, 219.

17 "Yes!" he wrote. "Well put." AG's copy of *Origin of Species* is housed at GH.

Chapter 3: Beetles, Birds, Theories

18 "two rare beetles" CD, *Autobiographies,* ed. Michael Neve and Sharon Messenger (New York: Penguin, 2002), 33.

18 "a disgrace to yourself" Ibid., 10.

20 "sailing in these latitudes" CD, *Journal of Researches into the Geology and Natural History of the Various Countries Visited by H.M.S. Beagle* (1839; reprint New York: Hafner, 1952), 190–91.

20 "the whole surface sparkled" Ibid., 191–93.

20 "I could not have believed" Ibid., 228.

21 "Mine is a bold theory" Adrian J. Desmond and James Richard Moore, *Darwin* (New York: W. W. Norton, 1994), 260.

21 "a thousand jasper steps" Erasmus Darwin, *The Temple of Nature, Or, The Origin of Society: A Poem, with Philosophical Notes* (Baltimore: John W. Butler, and Bonsal & Niles, 1804), 3–4, 71–72.

22 "inheritance of acquired characteristics" See Jean Baptiste Pierre Antoine de Monet de Lamarck, *Philosophie Zoologique: An Exposition with Regards to the Natural History of Animals,* trans. Hugh Elliot (1914; reprint Cambridge: Cambridge University Press, 2011).

22 **"the true and unmistakable head"** Robert Chambers, *Vestiges of the Natural History of Creation and Other Evolutionary Writings,* ed. James A. Secord (Chicago: University of Chicago Press, 1994), 272–73.

22 **"in our teeth, hands, and other features"** Ibid., 266.

25 **"the naturalist who accompanied"** *LAG,* 1:117.

25 **"The power of population"** Thomas R. Malthus, *An Essay on the Principle of Population* (Oxford: Oxford World's Classics, 2008), 13.

26 **"lively, agreeable person"** A. Hunter Dupree, *Asa Gray: American Botanist, Friend of Darwin* (Baltimore: Johns Hopkins University Press, 1988), 192.

26 **"As I am no Botanist"** CD, *The Life and Letters of Charles Darwin: Including an Autobiographical Chapter,* ed. Sir Francis Darwin (New York: Appleton, 1898), 60.

26 **"eminently glad to see your conclusion"** Ibid., 446.

26 **"Nineteen years (!) ago"** Ibid., 437.

27 **"fever it occurred to me that"** Alfred Russel Wallace, "The 'Ternate Paper,'" in *Infinite Tropics: An Alfred Russel Wallace Anthology,* ed. Andrew Berry (London: Verso, 2003), 51.

28 **"I never saw a more striking coincidence"** CD, *Life and Letters of Darwin,* 473.

Chapter 4: Word of Mouth

29 **"Eden in its primal beauty"** Michael Rawson, *Eden on the Charles: The Making of Boston* (Cambridge, Mass.: Harvard University Press, 2011), 164.

30 **"Boston always held her head too high"** Oliver Wendell Holmes, *The Complete Poetical Works* (Boston: Houghton Mifflin, 1908), 357.

30 **"Where Lowells talk only"** Joan Waugh, *Unsentimental Reformer: The Life of Josephine Shaw Lowell* (Cambridge, Mass.: Harvard University Press, 1997), 65.

30 **"no constitution or by-laws"** Holmes, *Complete Poetical Works,* 269.

31 **"cold & chilly as the gallery"** Charles Eliot Norton, *Letters* (Boston: Houghton Mifflin, 1913), 201.

32 **"a modified fish"** Ibid., 202.

32 **"crammed full of most interesting matter"** CD, *Correspondence,* 8:16.

33 **"When you *unscientific people*"** *LAG,* 2:462.

33 **"The South is our best customer"** Harold Holzer, *Lincoln at Cooper Union: The Speech that Made Abraham Lincoln President* (New York: Simon & Schuster, 2004), 67.

33 **"We are neither a Northern"** Hiram Ketchum, speech, *Journal of Commerce,* December 17, 1860.

34 **"perfect *flood* of humanity"** *LCLB,* 58–59.

34 **Many spoke little or no English** Allan Pred, *Urban Growth and City Systems in the United States, 1840–1860* (Cambridge, Mass.: Harvard University Press, 1980), 30.

34 **"squalid little wretches"** Lydia Maria Child, *Letters from New-York* (London: Richard Bentley, 1843), 62.

34 **"endless whirl of money-getting"** *LCLB,* 58–59.

35 **"Visited the Eleventh Ward"** "Charles Loring Brace, Diary 1853–55," n.p., n.d., NYHS.

35 **"prevention, not cure"** CLB, *The Dangerous Classes of New York and Twenty Years' Work Among Them* (New York: Wynkoop & Hallenbeck, 1872), 78.

35 **"More individuals are born"** CD, *Origin*, 467.

36 **"as fierce and bitter"** Charles Mackay, *Life and Liberty in America: Or, Sketches of a Tour in the United States and Canada, in 1857–8* (London: Smith, Elder & Co., 1859), 183.

36 **"bedlam of sounds"** *CLB, Dangerous Classes*, 194.

36 **"notorious rogues' den"** Ibid.

36 **"all the odds and ends"** *Ibid.*, 152.

37 **"not long for this world"** CLB, "Winter-Life Among the Poor," *Independent*, February 9, 1860, 6.

37 **"with lack-luster eye"** Ibid.

37 **"such thin pale faces"** Ibid.

37 **"daily and hourly scrutinising"** CD, *Origin*, 84.

37 **"houseless and frost-bitten"** CLB, "Winter-Life Among the Poor," 6.

38 **"the innocent victim"** CLB, "Children's Aid Society Seventh Annual Report," 3, NYHS.

38 **"In the distant future"** CD, *Origin*, 488.

40 **"become a convert"** Chauncey Wright, *Letters of Chauncey Wright; With Some Account of His Life*, ed. James Bradley Thayer (Boston: John Wilson, 1878), 43.

40 **"such ardor of praise"** Ibid., 30.

40 **"*Wyman*—the best of judges—"** CD, *Correspondence*, 8:27.

40 **"many disciples for Darwin"** William Barton Rogers, *Life and Letters of William Barton Rogers, Edited by His Wife* (Boston: Houghton Mifflin, 1896), 2:19.

41 **"John Brown's insurrectionary projects"** Ibid., 2:20.

41 **"such fairness and simplicity"** Untitled review of *Origin, Boston Courier,* March 5, 1860.

41 **"Darwinian before the letter"** Henry Adams, *Democracy / Esther / Mont Saint Michel and Chartres / The Education of Henry Adams* (New York: Library of America, 1983), 925.

41 **"nine men in ten"** Ibid., 926.

41 **"wrecking the Garden of Eden"** Ibid., 926.

42 **"The idea of violence"** Ibid., 758.

Chapter 5: Making a Stir

43 **"I believed myself already"** This anecdote is from Benjamin B. Hickok, *The Political and Literary Careers of F. B. Sanborn*, Ph.D. diss., Michigan State University, 1953, xi.

44 **"good crop of mystics"** Kenneth Sacks, *Understanding Emerson: "The American Scholar" and His Struggle for Self-Reliance* (Princeton, N.J.: Princeton University Press, 2003), 41.

44 **"wise wild beast"** FBS, *Transcendental and Literary New England: Emerson, Thoreau, Alcott, Bryant, Whittier, Lowell, Longfellow, and Others*, ed. Kenneth Walter Cameron (Hartford, Conn.: Transcendental Books, 1975), 22.

44 **"I am an ultra reformer"** D. R. Wilmes, "F. B. Sanborn and the Lost New England World of Transcendentalism," *Colby Library Quarterly* 16, no. 4 (December 1980): 240.

44 **"Frank Sanborn's little schoolhouse"** Julian Hawthorne, *The Memoirs of Julian Hawthorne* (New York: Macmillan, 1938), 85.

45 **"promised the greatest interest"** Madeleine Stern, *Louisa May Alcott: A Biography* (Lebanon, N.H.: University Press of New England, 1999), 80.

45 **"simple and conscientious master"** Hickok, *Political and Literary Careers of Sanborn*, 113.

46 **"piercing gray eyes"** FBS, *Recollections of Seventy Years* (Boston: R. G. Badger, 1909), 76.

46 **"a Calvinistic Puritan"** Ibid.

46 **"saw with unusual clearness"** Ibid., 77.

46 **"You see how it is"** FBS, *Table Talk: A Transcendentalist's Opinions on American Life, Literature, Art, and People from the Mid-Nineteenth Century Through the First Decade of the Twentieth*, ed. Kenneth Cameron (Hartford, Conn.: Transcendental Books, 1981), 143.

47 **"means to be on the ground"** FBS, *Recollections of Seventy Years*, 167.

47 **"annual chestnutting excursion"** and **"might compromise other persons"** Ibid., 187–88.

47 **"The whole matter was so uncertain"** Ibid., 188.

47 **"I shall refuse to obey"** Hickok, *Political and Literary Careers of Sanborn*, 171.

48 **"The Carolinian is widely different"** "South-Carolina," *New-England Magazine* 1, no. 3 (September 1831): 247.

48 ***"oran-outang,* or a South Sea Islander"** "James Russell Lowell and His Writings," *Debow's Review: Agricultural, Commercial, Industrial Progress and Resources* 28, no. 3 (September 1860): 273.

48 **"We are not one people"** Paul Calore, *The Causes of the Civil War: The Political, Cultural, Economic and Territorial Disputes Between North and South* (Jefferson, N.C.: McFarland, 2008), 257.

48 **"rival, hostile Peoples"** James M. McPherson, *Battle Cry of Freedom* (New York: Oxford University Press, 1988), 41.

48 **"Mr Brace brought a book"** FBS and Benjamin Smith Lyman, *Young Reporter of Concord: A Checklist of F. B. Sanborn's Letters to Benjamin Smith Lyman, 1853–1867, with Extracts Emphasizing Life and Literary Events in the World of Emerson, Thoreau and Alcott,* edited by Kenneth Cameron (Hartford, Conn.: Transcendental Books, 1978), 22.

49 **"no one who stands his ground"** Higginson to FBS, February 3, 1860, Boston Public Library.

49 **"is there no such thing"** Higginson to FBS, November 17, 1859, Boston Public Library.

50 **"I was not so much concerned"** Hickok, *Political and Literary Careers of Sanborn*, 177.

50 **"before I was quite ready"** Ibid., 206–7.

50 **"a thousand better ways"** Edward J. Renehan, Jr., *Secret Six: The True Tale of the Men Who Conspired with John Brown* (Columbia: University of South Carolina Press, 1997), 240.

Chapter 6: A Night at the Lyceum

51 **"disappearing like a hare"** Philip McFarland, *Hawthorne in Concord* (New York: Grove/Atlantic, 2007), 297.

52 **"drained my nerve power"** Richard Benson Sewall, *The Life of Emily Dickinson* (Cambridge, Mass.: Harvard University Press, 1994), 5.

52 **"a train of fifteen railroad cars"** John Matteson, *Eden's Outcasts: The Story of Louisa May Alcott and Her Father* (New York: W. W. Norton, 2007), 94.

52 **"the most eccentric man"** Charles Godfrey Leland, *Memoirs* (New York: D. Appleton & Co., 1893), 46.

53 **"Every soul feels"** ABA, "Orphic Sayings," *Dial* 1 (July 1840): 87.

53 **"Our age is retrospective"** RWE, "Nature," in *Essays and Lectures* (New York: Library of America, 1983), 7.

54 **"the top of our beings"** RWE, *Journals and Miscellaneous Notebooks,* ed. William H. Gilman et al. (Cambridge, Mass.: Belknap Press, 1975), 5:274.

54 **"currents of the Universal Being"** RWE, "Nature," in *Essays and Lectures,* 10.

54 **"The Baconian mode of discovery"** Francis Bowen, "Transcendentalism," *Christian Examiner and General Review* 21 (1837): 377.

55 **"the latest form of infidelity"** Andrews Norton, *A Discourse on the Latest Form of Infidelity: Delivered at the Request of the "Association of the Alumni of the Cambridge Theological School," on the 19th of July, 1839* (Cambridge, Mass.: J. Owen, 1839), 31.

55 **"a fuller *Revelation*"** Frederick C. Dahlstrand, *Amos Bronson Alcott: An Intellectual Biography* (Madison, N.J.: Fairleigh Dickinson University Press, 1982), 122.

55 **"the source of truth and virtue"** Ibid., 67.

55 **"presupposes his little pupils"** Megan Marshall, *The Peabody Sisters: Three Women Who Ignited American Romanticism* (Boston: Houghton Mifflin Harcourt, 2006), 315.

55 **A Boston lawyer was less charitable** Carlos Baker, *Emerson Among the Eccentrics: A Group Portrait* (New York: Viking, 1996), 184.

56 **"certain race of wild creatures"** Thomas Wentworth Higginson, *Out-Door Papers* (Boston: Lee & Shepard, 1886), 107.

56 **"Darwin, the naturalist, says"** *HDT,* 332–33.

57 **"society never advances"** RWE, "Self-Reliance," in *Essays and Lectures,* 279.

57 **"well-clad, reading, writing, thinking American"** Ibid.

57 **"latent distrust of civilization"** Higginson, *Out-Door Papers,* 108.

57 **"refinement and culture"** Ibid., 112.

57 **"always tending to decay"** Ibid., 115.

57 **"for the most degraded races"** Ibid., 108.

57 "takes more readily to civilization" Ibid., 116.

57 "so sad a tale" "The South and Free Negroes," *New York Times,* March 20, 1860, 4.

58 "a potent auxiliary" James A. Briggs and Abraham Lincoln, *An Authentic Account of Hon. Abraham Lincoln Being Invited to Give an Address: In Cooper Institute, N.Y., February 27, 1860* (Putnam, Conn.: Privately printed, 1915), n.p.

58 "a Dismal Swamp of inhumanity" Higginson, *Out-Door Papers,* 129.

58 "Thoreau's prejudice for Adamhood" Raymond R. Borst, *The Thoreau Log: A Documentary Life of Henry David Thoreau, 1817–1862* (Boston: G. K. Hall, 1992), 552.

59 "plodding through the prairie mud" RWE, *The Letters of Ralph Waldo Emerson,* ed. Eleanor Marguerite Tilton (New York: Columbia University Press, 1939), 5:194.

60 "a subtle chain" RWE, "Nature," in *Essays and Lectures,* 5.

60 "Nature is no sentimentalist" RWE, "The Conduct of Life," in *Essays and Lectures,* 945.

60 "The face of the planet" Ibid., 949.

61 "Naturalists . . . begin with matter" FBS and William Torrey Harris, *A. Bronson Alcott: His Life and Philosophy* (New York: Roberts Brothers, 1893), 2:486.

61 "It is like a swarm of bees" Ibid., 2:487.

61 "almost the *exact* truth" Ibid.

62 "Man's victory over nature" ABA, *The Journals of Bronson Alcott,* ed. Odell Shepard (New York: Little, Brown, 1938), 325.

62 "last of the philosophers" *HDT,* 535.

Chapter 7: The Nick of Time

64 "My good Henry Thoreau" RWE, *Journals and Miscellaneous Notebooks,* 4:453.

64 "'What are you doing now?'" HDT, *The Journal, vol. 1, 1837–1844,* ed. Elizabeth Witherell et al. (Princeton, N.J.: Princeton University Press, 1981), 5.

64 "an imitation of Emerson" Walter Roy Harding, *The Days of Henry Thoreau* (New York: Alfred A. Knopf, 1965), 299.

65 "Henry does not feel himself" Ibid., 301.

65 "As for taking Thoreau's arm" Kevin MacDonnell, "Collecting Henry David Thoreau," Walden Woods Project, http://bit.ly/1OLGYEq.

65 "another friendship is ended" HDT, *Journal,* 7:249.

65 "the education of every young man" RWE, "Man the Reformer," in *Essays and Lectures,* 139.

65 "live deep and suck out all the marrow" *HDT,* 394.

66 "to improve the nick of time" *HDT,* 336.

67 800 million pounds Sven Beckert, *Empire of Cotton: A Global History* (New York: Alfred A. Knopf, 2014), 243.

67 90 percent of all the slaves Ibid., 110.

68 nearly four million slaves Walter Johnson, *Soul by Soul: Life Inside the Antebellum Slave Market* (Cambridge, Mass.: Harvard University Press, 1999), 16, 2.

68 "You dare not make war" Frederick Law Olmsted, *The Cotton Kingdom: A Traveller's Observations on Cotton and Slavery in the American Slave States, 1853–1861,* ed. Arthur M. Schlesinger (1861; reprint New York: Da Capo Press, 1996), 7.

68 "If an angry bigot assumes" RWE, "Self-Reliance," in *Essays and Lectures,* 262.

68 "I will not obey it" RWE, *Emerson's Antislavery Writings,* ed. Len Gougeon and Joel Myerson (New Haven, Conn.: Yale University Press, 2002), xxxviii.

69 "keep a *cordon sanitaire*" Randall Fuller, *From Battlefields Rising: How the Civil War Transformed American Literature* (New York: Oxford University Press, 2011), 57.

69 "I think we must get rid of slavery" Ibid.

69 "We have attempted" RWE, *The Collected Works of Ralph Waldo Emerson,* ed. Ronald A. Bosco and Douglas Emory Wilson (Cambridge, Mass.: Belknap Press, 2007), 7:332.

69 "I cannot for an instance" *HDT-CE,* 206.

69 the John Brown affair *The Oxford Handbook of Transcendentalism,* ed. Joel Myerson, Sandra Harbert Petrulionis, and Laura Dassow Walls (New York: Oxford University Press, 2010), 218.

70 "I did not send to you" Harding, *Days of Thoreau,* 417.

70 as if it "burned him" Edward Waldo Emerson, *Henry Thoreau: As Remembered by a Young Friend* (Boston: Houghton Mifflin, 1917), 71.

70 "a revolutionary Lecture" ABA, *The Journals of Bronson Alcott,* ed. Odell Shepard (New York: Little, Brown, 1938), 2:384.

70 "mightily stirred by the emotions" FBS, *The Personality of Thoreau* (Boston: C.E. Goodspeed, 1901), 36–37.

70 "a thorough fanatic" Sandra Harbert Petrulionis, *To Set This World Right: The Antislavery Movement in Thoreau's Concord* (Ithaca, N.Y.: Cornell University Press, 2006), 137–38.

71 "it was more fitting to signify" Ibid., 139.

71 "So universal and widely related" Ibid.

71–72 "young man is a demigod" HDT, *Journal,* 13:35.

72 "the character of my knowledge" Harding, *Days of Thoreau,* 291.

72 "What sort of science" HDT, *Journal,* 3:156.

72 "I awoke this morning" Ibid., 3:150.

72 "His theory of the formation" Ibid., 2:263.

73 a small diary and a pencil FBS, *Henry D. Thoreau* (Boston: Houghton Mifflin, 1910), 251.

73 "I am drawing a rather long bow" HDT, *Journal,* 6:411.

74 "a hungry, omnivorous monster" John Burroughs, *The Late Harvest* (Boston: Houghton Mifflin, 1922), 148.

74 "Observation is so wide awake" *HDT,* 296.

74 "choice documents to me" Frederick Douglass, *Autobiographies: Narrative of the Life of Frederick Douglass, an American Slave / My Bondage and My Freedom / Life and Times of Frederick Douglass,* ed. Henry Louis Gates (New York: Library of America, 1994), 226.

75 **"I have indulged myself"** Barbara Hochman, *"Uncle Tom's Cabin" and the Reading Revolution: Race, Literacy, Childhood, and Fiction, 1851–1911* (Amherst: University of Massachusetts Press, 2011), 84.

Chapter 8: Bones of Contention

80 **invited to be a corresponding member in 1850** William Ellery Channing, *Thoreau, the Poet-Naturalist, with Memorial Verses,* ed. FBS (Boston: Charles E. Goodspeed, 1902), 281.

80 **"Gorilla collection of Mr. Du Chaillu"** "January 4, 1860," *Proceedings of the Boston Society of Natural History* 7 (1861), 211.

81 **"one of the troglodyte tribe"** Monte Reel, *Between Man and Beast: An Unlikely Explorer and the African Adventure That Took the Victorian World by Storm* (New York: Anchor, 2013), 91.

81 **"the teeth were in a continuous series"** "January 4, 1860," *Boston Society of Natural History* 7 (1861), 212.

82 **"the relative size of the brain"** Ibid.

82 **"In the young gorilla"** Ibid., 211–12.

83 **"Take with this their awful cry"** Paul du Chaillu, *Explorations and Adventures in Equatorial Africa* (New York: Harper & Brothers, 1861), 126.

83 **"*poor—very poor*!!"** CD, *Correspondence,* 8:16.

83 **"the splendid, magnificent letter"** Ibid., 8:45.

Chapter 9: Agassiz

85 **"of men, skulls, skeletons"** "December 21, 1859," *Proceedings of the Boston Society of Natural History* 7 (1861), 192.

85 **"the free conception of the Almighty Intellect"** Guy Davenport, *The Intelligence of Louis Agassiz: A Specimen Book of Scientific Writings* (Boston: Beacon Press, 1963), ix.

86 **"There is order in the universe"** AG, *Darwiniana: Essays and Reviews Pertaining to Darwinism* (Cambridge, UK: Cambridge University Press, 2009), 58.

86 **"behind nature, throughout nature"** RWE, "Nature," in *Essays and Lectures,* 7.

86 **"study of nature"** Charles Frederick Holder, *Louis Agassiz: His Life and Work* (New York: Putnam's 1893), 193.

86 **"He is a fine, pleasant fellow"** *LAG,* 343.

86 **"a good genial fellow"** A. Hunter Dupree, *Asa Gray: American Botanist, Friend of Darwin* (Baltimore: Johns Hopkins University Press, 1988), 228.

87 **"writing and talking *ad populum*"** Louis Menand, *The Metaphysical Club: A Story of Ideas in America* (New York: Macmillan, 2001), 125.

88 **"In surveying the globe"** "The Natural History of Man," *Methodist Review* 26 (1844): 257.

88 **"hardly be dated"** "Mr. Brace on the Races of the Old World," *New York Times,* June 22, 1863.

89 "the likeness of the Creator" James Cowles Prichard, *Researches Into the Physical History of Man* (London: Arch, 1813), 1.

89 "constitute but one race" Ibid., iii.

89 "on a race originally uniform" Ibid., 2.

89 "primitive stock of men" Ibid., 233.

90 "We have no great objection" Charles Caldwell, *Thoughts on the Original Unity of the Human Race* (Cincinnati: J.A. & U.P. James, 1852), 34.

90 "to the male ape" Ibid., 57.

91 "Look first upon the Caucasian female" Josiah Nott, "American Intelligence," *American Journal of the Medical Sciences* 5 (1843): 254.

91 "Is that a fair objection" Stephen Jay Gould, *The Mismeasure of Man* (New York: W. W. Norton, 2006), 77.

91 "resolved to fight" Frederick Douglass, *Autobiographies; Narrative of the Life of Frederick Douglass, an American Slave / My Bondage and My Freedom / Life and Times of Frederick Douglass*, ed. Henry Louis Gates (New York: Library of America, 1994), 64–65.

91 "turning point in my career as a slave" Ibid.

92 "the vital question of the age" Frederick Douglass, *The Claims of the Negro, Ethnologically Considered: An Address, Before the Literary Societies of Western Reserve College, at Commencement, July 12, 1854* (Rochester, N.Y.: Lee, Mann, 1854), 5.

92 "BECAUSE HE IS NOT A MAN!" Ibid., 7.

92 "no human, barely any animal feeling" Tania Das Gupta et al., *Race and Racialization: Essential Readings* (Toronto: Canadian Scholars' Press, 2007), 26.

92 "a phalanx of learned men" Douglass, *Claims of the Negro*, 285.

92 "the Notts, the Gliddens [*sic*], the Agassiz" Ibid., 10.

92 "most glory upon the wisdom" Ibid., 11.

92 "The unity of the human race" Ibid., 12.

93 "at the bottom of the whole controversy" Ibid., 13.

93 "separate the negro race" Ibid., 16.

93 "staked out the ground beforehand" Ibid., 20.

93 "endeavor to make you" David Brion Davis, *The Problem of Slavery in the Age of Emancipation* (New York: Alfred A. Knopf, 2014), 38.

93 "the nakedness of some [slaves]" Ibid., 203–4.

94 "A slave is *not*" Ibid., 270, my emphasis.

94 "[Is it] not a little remarkable" J. C. Nott and George R. Gliddon, *Types of Mankind Or Ethnological Researches, Based Upon the Ancient Monuments, Paintings, Sculptures, and Crania of Races, and Upon Their Natural, Geographical, Philological, and Biblical History* (Philadelphia: Lippincott, 1857), 75.

94 "Man is distinguished" Douglass, *Claims of the Negro*, 8.

95 "a certain class of ethnologists" John Stauffer et al., *Picturing Frederick Douglass: An Illustrated Biography of the Nineteenth Century's Most Photographed American* (London: Liveright, 2015), 167.

95 "all the rest intermediates!" Douglass, *Claims of the Negro*, 8.

Chapter 10: The What-Is-It?

96 "even in a country store" Christopher Irmscher, *The Poetics of Natural History: From John Bartram to William James.* (New Brunswick, N.J.: Rutgers University Press, 1999), 104.

96–97 "My organ of acquisitiveness" P. T. Barnum, *The Life of P. T. Barnum, Written by Himself* (London: Sampson Low, 1855), 20.

97 "promoters of the natural sciences" Irmscher, *Poetics of Natural History,* 103.

97 "the gods visible again" Ibid., 300.

97 "A man has been found" Bluford Adams, *E Pluribus Barnum: The Great Showman and the Making of U.S. Popular Culture* (Minneapolis: University of Minnesota Press, 1997), 148.

98 "we wage war" Sidney Kaplan, *American Studies in Black and White: Selected Essays, 1949–1989* (Amherst: University of Massachusetts Press, 1996), 200.

98 "Most Marvelous Creature Living" Irmscher, *Poetics of Natural History,* 132.

98 "what manure is to the land" P. T. Barnum, *The Humbugs of the World: An Account of Humbugs, Delusions, Impositions, Quackeries, Deceits, and Deceivers Generally, in All Ages* (New York: Carleton, 1866), 66.

99 "He possesses the skull, limbs" Irmscher, *Poetics of Natural History,* 132.

99 "apparently more strength" Ibid., 135.

99 "he's a great fact for Darwin" George Templeton Strong, *The Diary,* ed. Allan Nevins (New York: Macmillan, 1952), 12.

99 "to see the much advertised" Ibid.

99 "Darwin cannot understand" Ibid., 13.

100 "got hold of *a* truth" Ibid., 10.

100 "certain elemental atoms" Ibid., 14.

100 "thousands of millions" Ibid., 10.

100 "flying fish by successive minute steps" Ibid., 11.

100 "some ancestral archaic fish" Ibid., 10.

101 "The showman's story" Ibid., 12.

101 "a grand grizzly bear" Ibid.

102 "Well, we fooled 'em" Bernth Lindfors, *Early African Entertainments Abroad: From the Hottentot Venus to Africa's First Olympians* (Madison: University of Wisconsin Press, 2014), 170.

102 "modified or changed" CLB, *The Races of the Old World: A Manual of Ethnology* (London: John Murray, 1863), 147.

103 "I am anxious to talk" *LCLB,* 210.

103 "deepest feeling of my heart" Ibid., 72.

103 "The shadow of our national sin" Ibid., 256.

103 "Evil seems to me destructive" Ibid., 335.

103 "The idea of this age" Ibid., 297.

104 "I think it furnishes what historians" Ibid., 285.

104 "We all know how energetic" "Correspondence from Boston," *Daily National Intelligencer,* February 20, 1860, col. C.

105 **"kept screws to crush the fingers"** CD, *The Voyage of the Beagle: Charles Darwin's Journal of Researches,* ed. Janet Browne and Michael Neve (London: Penguin Classics, 1989), 499.

105 **"the vilest of the human kind"** Adrian Desmond and James Moore, *Darwin's Sacred Cause: How a Hatred of Slavery Shaped Darwin's Views on Human Evolution* (Boston: Houghton Mifflin Harcourt, 2009), 10. This is by for the best account of Darwin's lifelong aversion to New World slavery.

105 **"I shall never forget my feelings"** CD, *Voyage,* 28.

105 **"heard the most pitiable moan"** Desmond and Moore, *Darwin's Sacred Cause,* 182.

Chapter 11: A Spirited Conflict

107 **"we thought Darwin had thrown"** James McIntosh notes, "In all likelihood Dickinson read the series of excellent, informative articles by Asa Gray." McIntosh, *Nimble Believing: Dickinson and the Unknown* (Ann Arbor: University of Michigan Press, 2000), 174n16; *Letters of Emily Dickinson,* ed. Thomas H. Johnson (1958; reprint Cambridge, Mass.: Belknap Press, 1997), letter 750.

108 **"sneer at the idea"** "Christian Theism: The Testimony of Reason and Revelation to the Existence and Character of the Supreme Being," *New Englander and Yale Review* 14, no. 56 (1856): 628. The remaining statements against Darwin's theory are found in AG, "Darwin and His Reviewers," *Atlantic Monthly* 6 (October 1860), 409, 411.

108 **"a *universal and ultimate* physical cause"** AG, *Darwiniana: Essays and Reviews Pertaining to Darwinism* (Cambridge: Cambridge University Press, 2009), 137.

108 **"is inconsistent with the idea"** Ibid., 54.

108 **"independent, specific creation"** Ibid., 10–11.

109 **"theistic to excess"** Ibid., 14.

109 **"the domain of inductive science"** Ibid.

109 **"subject from their birth to *physical* influences"** Ibid., emphasis added.

109 **"A spirited conflict"** Ibid., 10.

110 **"As the conclusions"** CD, *The Life and Letters of Charles Darwin: Including an Autobiographical Chapter,* ed. Sir Francis Darwin (New York: Appleton, 1898), 11.

110 **"A sentence likely to mislead!"** Louis Agassiz's copy of *On the Origin of Species,* 187, 184, MCZ.

110 **"The mistake of Darwin"** Ibid., 112.

111 **"does the excellences of the classification"** Ibid., 129.

111 **"Why sir, there is room"** Ibid., 186.

111 **"There is a design"** Agassiz quoted in CD, *Correspondence,* 8:55.

111 **"Tell Darwin that Agassiz"** Edward Lurie, *Louis Agassiz: A Life in Science* (Chicago: University of Chicago Press, 1960), 295.

113 **"the most degraded of human races"** Stephen Jay Gould, *The Mismeasure of Man* (New York: W. W. Norton, 2006), 36.

113 **"feeling that they inspired in me"** Ibid., 44–45.

114 **"more widely different"** Elizabeth Cabot Cary Agassiz, *Louis Agassiz: His Life and Correspondence* (Boston: Houghton, Mifflin, 1893), 603.

114 **"one of the most difficult problems"** Ibid.

115 **"best marked human races"** AG, *Darwiniana,* 51.

Chapter 12: Into the Vortex

117 **"Every few weeks"** LMA, *Little Women: Or, Meg, Jo, Beth, and Amy* (1868–69; reprint New York: Little, Brown 1922), 214.

118 **"There blossomed forth"** James Redpath, *Echoes of Harper's Ferry* (Boston: Thayer & Eldridge, 1860), 98.

118 **"set my teeth & vowed"** LMA, *The Selected Letters of Louisa May Alcott,* ed. Joel Myerson and Daniel Shealy (Athens: University of Georgia Press, 1995), 34.

119 **Orphans populate American fiction** See, for instance, Carol J. Singley, *Adopting America: Childhood, Kinship, and National Identity in Literature* (New York: Oxford University Press, 2011).

119 **"the man who has helped me"** John Matteson, *Eden's Outcasts: The Story of Louisa May Alcott and Her Father* (New York: W. W. Norton, 2007), 407.

120 **"encourage and lead her"** ABA, *The Journals of Bronson Alcott,* ed. Odell Shepard (New York: Little, Brown, 1938), 326.

120 **"Though in many people's eyes"** Katharine Susan Anthony, *Louisa May Alcott* (Westport, Conn.: Greenwood Press, 1977), 109.

120 **"I am pleased"** ABA, *Journals of Alcott,* 326.

121 **"They are bright girls"** FBS to Theodore Parker, February 12, 1860, CFPL.

121 **"This court acknowledges"** John Stauffer and Zoe Trodd, *The Tribunal: Responses to John Brown and the Harpers Ferry Raid* (Cambridge, Mass.: Harvard University Press, 2012), 55.

122 **"Ah, he is too blue"** Anne Brown Adams, "Louisa May Alcott in the Early 1860s," in *Alcott in Her Own Time: A Biographical Chronicle of Her Life, Drawn from Recollections, Interviews, & Memoirs by Family, Friends, & Associates,* ed. Daniel Shealy (Iowa City: University of Iowa Press, 2005), 11.

122 **"armed with tar"** William G. Allen, *American Prejudice Against Color; An Authentic Narrative, Showing How Easily the Nation Got into an Uproar* (London: W. & F. G. Cash, 1853), 1.

123 **"a Spaniard, and of noble family"** LMA, *Louisa May Alcott on Race, Sex, and Slavery,* ed. Sarah Elbert (Boston: Northeastern University Press, 1997), 53–54.

124 **"odours of orange-flowers"** Henry Wadsworth Longfellow, *Longfellow's Poetical Works: With 83 Illustrations by Sir John Gilbert, R.A., and Other Artists* (London: George Routledge & Sons, 1883), 30.

124 **a hint of humid sexuality** LMA, *Alcott on Race, Sex, and Slavery,* 68.

124 **"lifted up into humanity"** Ibid., 18.

125 **"true-blue May"** Ibid., xvi.

125 **"very nice of Mr. Swedenborg"** Frederick C. Dahlstrand, *Amos Bronson Alcott: An Intellectual Biography* (Madison, N.J.: Fairleigh Dickinson University Press, 1982), 232.

125 **"general evil effects"** Abraham Lincoln, "Address on Colonization to a Deputation of Negroes," August 14, 1862, http://bit.ly/1Qhsj6z

125 **"Diviner love"** LMA, *Alcott on Race, Sex, and Slavery,* 72–76.

126 **"a welcome to that brotherhood"** Ibid., 27.

126 **"the weaker now"** Ibid., 22–24.

126 **"Mr——won't have"** LMA, *The Journals of Louisa May Alcott,* ed. Joel Myerson, Daniel Shealy, and Madeleine B. Stern (Athens: University of Georgia Press, 1997), 98.

127 **"organized on the basis of making war"** David Goldfield, *America Aflame: How the Civil War Created a Nation* (New York: Bloomsbury Press, 2011), 161.

Chapter 13: Tree of Life

128 **"I have sent it to Agassiz"** CD, *Correspondence,* 8:95.

128 **"by far the most able"** Ibid., 8:106.

128 **"You have succeeded"** A. Hunter Dupree, *Asa Gray: American Botanist, Friend of Darwin* (Baltimore: Johns Hopkins University Press, 1988), 277.

128 **"the best yet on the subject"** Ibid., 278.

129 **Investigations into the way** AG, *Darwiniana: Essays and Reviews Pertaining to Darwinism* (Cambridge: Cambridge University Press, 2009), 87.

129 **"no good to old beliefs"** Ibid., 87–88.

129 **"the new doctrine was better"** Ibid., 87.

129 **"when we consider"** Ibid., 89–92.

130 **"represented by a great tree"** CD, *Origin,* 129.

130 **"all living things have much in common"** Ibid., 484.

131 **"together all languages"** Ibid., 422–23.

131 **"just as philologists"** AG, *Darwiniana,* 98.

132 **"disfigured with smoke"** CLB, "Charles Loring Brace, Diary 1853–55," 176, NYHS.

132 **"Last night, I slept"** Ibid., 175.

132 **"The trial of fifty years"** Mason I. Lowance, *Against Slavery: An Abolitionist Reader* (New York: Penguin, 2000), 245.

133 **"Slavery is a** *sin per se*" LCLB, 67.

133 **"silently and insensibly working"** CD, *Origin,* 84, my emphasis.

133 **"such casual criticism"** *LAG,* 461.

134 **"cut off all future immigration"** Ibid., 462.

134 **"There is danger"** Charles Eliot Norton to AG, June 22, 1860, GH.

Chapter 14: A Jolt of Recognition

136 **Thoreau copied the sentence** All transcriptions are from HDT's "Extracts Mostly upon Natural History," holograph notes, unsigned and undated, NYPL.

137 **"could tell accurately"** *HDT,* 551.

137 **"all contained in a breakfast cup!"** CD, *Origin,* 386–87.

137 **"Lying between the earth"** *HDT,* 463.

137 **"From a hill-top"** Ibid., 472.

138 **"the whole economy of nature"** CD, *Origin,* 62.

138 "I expected a fauna" HDT, *The Heart of Thoreau's Journals,* ed. Odell Shepard (Boston: Houghton Mifflin, 1927), 198.

138 "How simply is the fact explained" CD, *Origin,* 473.

139 "Why do precisely these objects" *HDT,* 502.

139 "We can never arrive" Alfred Russel Wallace, *Island Life or the Phenomena and Causes of Insular Faunas and Floras, Including a Revision and Attempted Solution of the Problem of Geological Climates* (1880; reprint New York: Macmillan, 1902), 13.

141 "I have not found" CD, *Origin,* 393.

141 "[The French naturalist]" Ibid.

141 "I love to see" *HDT,* 576.

141 "in the hollow" Ibid., 575–76.

142 "Every one has heard" CD, *Origin,* 74–75.

142 "Throw up a handful" Ibid., 75.

142 "Innumerable little streams" *HDT,* 565.

143 "stood in the laboratory" Ibid., 566.

143 a primordial "ur-plant" Laura Dassow Walls, *Seeing New Worlds: Henry David Thoreau and Nineteenth-Century Science* (Madison: University of Wisconsin Press, 1995), 35.

143 "the original forms of vegetation" *HDT,* 566.

143 "What is man" Ibid., 567–68.

144 "It is interesting to contemplate" CD, *Origin,* 489–90.

145 "leaves of goldenrod" HDT, "Nature Notes, Charts and Tables: Autograph Manuscript," n.p., NYPL.

Chapter 15: Wildfires

148 "We who studied" Moncure Daniel Conway, *Autobiography* (London: Cassell, 1904), 1:249.

148 "successful writer and natural historian" "February 15, 1860," *Proceedings of the Boston Society of Natural History* 7 (1859–61), 231.

148 "Animal representatives were as numerous" Ibid.

149 "made some little fling" Conway, *Autobiography,* 1:152–53.

150 "Any faith declaring" ABA, *The Journals of Amos Bronson Alcott,* ed. Odell Shepard (New York: Little, Brown, 1938), 680.

150 "grub state" RWE, "The American Scholar," in *Essays and Lectures,* 61.

152 "He is now convinced" Ronald A. Bosco and Jillmarie Murphy, *Hawthorne in His Own Time: A Biographical Chronicle of His Life* (Iowa City: University of Iowa Press, 2007), 104.

152 "a celebration of the principles" ABA, *Diary,* 50.

153 "'a kind of bow-arrow tang'" *HDT-CE,* 458.

153 "Who knows but like the dog" HDT, "Wild Apples Manuscript," n.p., NYPL.

153 "a web of complex relations" CD, *Origin,* 73.

153 "Many cases are on record" Ibid., 71.

154 "the introduction of a single tree" Ibid.

155 "it was beyond . . . reach" HDT, *Journal,* 8:22.

156 "the dangerous time" Ibid., 13:236.

157 "Science in many departments" Ibid., 13:154.

158 "The man's fellow-laborer" Ibid., 13:40–41.

Chapter 16: Discord in Concord

162 "the dirty planet" Joan W. Goodwin, *The Remarkable Mrs. Ripley: The Life of Sarah Alden Bradford Ripley* (Boston: Northeastern University Press, 2011), 115, 116.

162 "increase of population" Sean Wilentz, ed., *The Best American History Essays on Lincoln* (New York: Macmillan, 2008), 49.

163 "as coolly as Macaulay" Reprinted in *Boston Daily Advertiser,* March 16, 1860, col. F.

163 "When the struggle" Abraham Lincoln, *Speeches of Abraham Lincoln: Including Inaugurals and Proclamations* (New York: A. L. Burt, 1906), 290.

163 "blowing out the moral lights" Abraham Lincoln, "Speech at Columbus, Ohio," September 16, 1859, in *Complete Works, Comprising His Speeches, State Papers, and Miscellaneous Writings,* ed. John G. Nicolay and John Hay (New York: Century, 1920), 557.

164 "Old John Brown" Michael Burlingame, *Abraham Lincoln: A Life* (Baltimore, Md.: Johns Hopkins University Press, 2013), 566.

164 "to destroy the Union" Ibid., 576.

164 Sanborn supplemented Walker's history FBS to Theodore Parker, March 11, 1860, CFPL.

165 "has just been reading Darwin's" Elizabeth Hoar, *Mrs. Samuel Ripley* (New York: Lippincott, 1877), 89.

167 "encompassed by a throng of men" Walter Roy Harding, *The Days of Henry Thoreau* (New York: Alfred A. Knopf, 1965), 423.

167 "Have you ever enjoyed" Ellen Tucker Emerson, *The Letters of Ellen Tucker Emerson* (Kent, Ohio: Kent State University Press, 1983), 1:212–13.

167 "a new sort of amusement" LMA, *The Journals of Louisa May Alcott,* ed. Joel Myerson, Daniel Shealy, and Madeleine B. Stern (Athens: University of Georgia Press, 1997), 53.

168 "committed to the custody" FBS, *Recollections of Seventy Years* (Boston: R. G. Badger, 1909), 211.

168 "Where are we?" "The Outrage at Concord," *Independent,* April 12, 1860.

168 "I always hold Sanborn" Horace Traubel, *With Walt Whitman in Camden* (Boston: Small, Maynard, 1906), 1:285.

168 "the issue was met at Concord yesterday" Benjamin Hickok, *The Political and Literary Careers of F. B. Sanborn,* Ph.D. diss., Michigan State University, 1953, 188.

169 "against any Senate's office" FBS to Theodore Parker, April 1, 1860, CFPL.

169 "heard the bells ringing" "Sanborn Arrest," *New York Herald,* April 7, 1860.

170–171 "What we are coming to" Samuel May to Abigail Alcott, April 16, 1860, HL.

171 "Ours is the property invaded" *Proceedings of the Conventions at Charleston and Baltimore, Published by the Order of the National Democratic Convention,*

(*Maryland Institute, Baltimore*), *and Under the Supervision of the National Democratic Executive Committee.* (Washington, D.C., 1860), 69.

Chapter 17: Moods

172 **"all grace and becomingness"** John Matteson, *Eden's Outcasts: The Story of Louisa May Alcott and Her Father* (New York: W. W. Norton, 2007), 255.

172 **"I mourn the loss"** LMA, *The Journals of Louisa May Alcott*, ed. Joel Myerson, Daniel Shealy, and Madeleine B. Stern (Athens: University of Georgia Press, 1997), 99.

173 **"a tall, stout woman"** LMA to Anna Alcott Pratt, May 27, 1860, in *The Selected Letters of Louisa May Alcott,* ed. Joel Myerson and Daniel Shealy (Athens: University of Georgia Press, 1995), 55.

173 **"Mr. H. is as queer as ever"** Ibid., 57.

174 **"standing betwixt man and animal"** Nathaniel Hawthorne, *The Marble Faun,* (London: Everyman, 1995), 13.

174 **"If you consider him well"** Ibid., 17.

174 **"It is a wonderful coincidence"** Frank Preston Stearns, *The Life and Genius of Nathaniel Hawthorne* (J. B. Lippincott, 1906), 368–69.

175 **"made a long stride"** Ibid., 369.

175 **"really a rise in life"** Ibid., 380.

175 **"Genius burned so fiercely"** LMA, *Journals of Louisa Alcott,* 99.

176 **"The husband's charm"** LMA, *Moods,* ed. Sarah Elbert (New Brunswick, N.J.: Rutgers University Press, 1991), 161.

176 **"a spiritual slavery"** Ibid., 8.

177 **"the pestilence of slavery"** Ibid., 12.

177 **"she received pride"** Ibid., 84.

177 **"Many acres were burning"** Ibid., 39.

178 **"Shy thing! I will tame you"** Ibid., 127.

179 **"subtler than perception"** Ibid., 135.

179 **"The trouble," Mandelet observes** Kate Chopin, *The Awakening* (New York: W. W. Norton, 1994), 105.

180 **"This peaceful mood"** LMA, *Moods,* 203.

180 **"Mr. Emerson offered to read it"** LMA, *Journals of Louisa Alcott,* 99.

Chapter 18: Meditations in a Garden

181 **"restless, domineering devil"** LMA, *Moods,* ed. Sarah Elbert (New Brunswick, N.J.: Rutgers University Press, 1991), 203.

181 **"Resistance to wrong"** Dwight Lowell Dumond, *Southern Editorials on Secession* (Gloucester, Mass.: Peter Smith, 1964), 140.

182 **"And for the Union"** Ralph L. Rusk, *The Letters of Ralph Waldo Emerson* (New York: Columbia University Press, 1939), 5:18.

182 **"My motto has long"** Sandra Harbert Petrulionis, *To Set This World Right: The Antislavery Movement in Thoreau's Concord* (Ithaca, N.Y.: Cornell University Press, 2006), 153.

183 "souls are *alike* essentially" Frederick C. Dahlstrand, *Amos Bronson Alcott: An Intellectual Biography* (Madison, N.J.: Fairleigh Dickinson University Press, 1982), 271.

183 "Individualism is brute" ABA, *Diary* (1861), 58.

183 "similar and partakers" Sidney H. Morse and Joseph B. Marvin, *The Radical* (Boston: A. Williams & Co., 1869), 23.

184 "Our thoughts are the offspring" Dahlstrand, *Amos Bronson Alcott*, 67.

184 "Truth and love" Ibid., 68.

184 "The Progress of Mankind" ABA, *Diary* (1860), n.p., HL.

185 "constitutes the ideal principle" Ibid.

185 "We belong now" Harry de Puy, "Amos Bronson Alcott: Natural Resource or Consecrated Crank?" *American Transcendental Quarterly* 1 (March 1987): 53.

186 "Now comes Darwin" Morse and Marvin, *Radical*, 23.

186 "Conway's Dial has some good things" FSB to Theodore Parker, March 11, 1860, CFPL.

186 "The awkward, the monstrous" "Review, Darwin's Origin of Species," *Dial* 1 (March 1860): 196–97.

187 "however profound our ignorance" M.B.B., "Darwin's Origin of Species," *Dial* 1 (June 1860): 391–92.

187 "The highest types of beauty" Ibid.

187 "Mr. Darwin's theory" Ibid.

188 "How shall we account" Ibid.

189 "In the year (1836)" Moncure Daniel Conway, *Autobiography* (London: Cassell, 1904), 1:360.

189 "the way upward" Ibid., 1:168.

189 "Our popular Christianity" Edwin C. Walker, *A Sketch and an Appreciation of Moncure Daniel Conway* (New York: Walker, 1908), 25.

189 "Evening: I am at Emerson's" ABA, *Diary* (1860), 269.

Chapter 19: The Succession of Forest Trees

190 "under rather unfavorable auspices" "Middlesex Agricultural Series II," CFPL.

191 "on Nature's Methods of planting" ABA, *Diary* (1860), 311–12.

191 "Everyone has heard" CD, *Origin,* 74–75.

191 "Professor Agassiz corroborated" A. Hunter Dupree, *Asa Gray: American Botanist, Friend of Darwin* (Baltimore: Johns Hopkins University Press, 1988), 257.

192 "Amid the acclamations" "The Capture and Occupation of Richmond," *Papers of Military Historical Society of Massachusetts* 14 (1918): 138.

192 "a coarse, yellow, sandy soil" Frederick Law Olmsted, *The Cotton Kingdom: A Traveller's Observations on Cotton and Slavery in the American Slave States, 1853–1861,* ed. Arthur M. Schlesinger (1861; reprint New York: Da Capo Press, 1996), 52.

192 "I affirmed . . . confidently years ago" *HDT-CE*, 433.

192 "On the 24th of September" Ibid., 433–34.

193 "Nature can persuade us" Ibid., 432.

193 "There is a patent office" Ibid., 431.

193 "laws ordained by God" CD, *Correspondence*, 8:224.

194 "long extinct plants" *HDT-CE*, 442.

195 "perfect alchemists I keep" Ibid., 443.

195 "Yet farmers' sons" Ibid.

Chapter 20: Races of the Old World

196 "Friend Thoreau" Bradley P. Dean, "Henry D. Thoreau and Horace Greeley Exchange Letters on the 'Spontaneous Generation of Plants,'" *New England Quarterly* 66 (December 1993): 633–34.

198 "solid basis of facts" CLB, "Ethnological Fallacies," *Independent*, December 20, 1860, 4.

198 "most severe between the individuals" CD, *Origin*, 75.

198 "A barbarous race" CLB, "Ethnological Fallacies."

199 "There is nothing" CLB, *The Races of the Old World: A Manual of Ethnology* (London: John Murray, 1863), 310.

199 "one of the saddest spectacles" Ibid., 351.

200 "so degenerated that they have abandoned" Ibid., 475.

200 "two centuries of degradation" Ibid., 477.

200 "The African peoples" Ibid., 312.

200 "free-born negro children" Ibid., 477–78.

200 "a succession of types" Ibid., 460.

200 "Scarcely any marks" Ibid., 505.

201 "due not merely to elevation" Ibid., 234.

201 "color and physical traits" Ibid., 152.

201 "Each parent is adapted" Ibid., 490.

201 "the people of the Cotton States" "The Pine and Palm," *Boston and New York Pine and Palm*, May 18, 1861, http://bit.ly/20hFjxL

202 "negro present[ed] his pure type" CLB, *Races of the Old World*, 502.

202 "There is every reason" Ibid., 492.

203 "more widely different" Elizabeth Cabot Cary Agassiz, *Louis Agassiz: His Life and Letters* (Boston: Houghton Mifflin, 1893), 595.

203 "less than two million blacks" Louis Menand, *The Metaphysical Club: A Story of Ideas in America* (New York: Macmillan, 2001), 114.

203 "the great design" CLB, *Races of the Old World*, 512–13.

Chapter 21: A Cold Shudder

204 he had hurled himself This description of Darwin's activities is derived from Janet Browne, *Charles Darwin: The Power of Place* (New York: Alfred A. Knopf, 2003), 2:206.

205 "Who shall number" CD, *Correspondence*, 8:160n5.

206 "The Lord hath delivered him" Keith Thomson, "Huxley, Wilberforce and the Oxford Museum," *American Scientist*, May-June 2000, http://bit.ly/1PJd17H

206 "fighting like a Trojan" CD, *Correspondence*, 8:280.

206 "'Almost thou persuadest me'" A. Hunter Dupree, *Asa Gray: American Botanist, Friend of Darwin* (Baltimore: Johns Hopkins University Press, 1988), 299.

207 "made a mistake in being a Botanist" CD, *Correspondence,* 8:299.

207 "that specific creation" Charles Eliot Norton to Asa Gray, June 22, 1860, GH.

207 "we must needs believe" AG, *Darwiniana: Essays and Reviews Pertaining to Darwinism* (Cambridge: Cambridge University Press, 2009), 92.

207 "I wish that you would give" Charles Eliot Norton to AG, June 22, 1860, GH.

207 "It does not seem probable" Ibid.

208 "of Gray as well as of Darwin" Ibid.

208 "inquire after the motives" AG, *Darwiniana,* 106–7.

208 "Why," he asked, "should a theory" Ibid., 107–09.

208 Why did bird and mammal Ibid., 121–22.

208 "and then . . . unequivocally vegetable" Ibid., 124.

209 "*an extraordinary degree of care*" William Paley, *Paley's Natural Theology* (London: Charles Knight, 1836), 1:42.

209 "reason tells me" CD, *Origin,* 186–87.

210 "So it does" AG's copy of *Origin,* 186, GH.

210 "the weakest point in the book" CD, *Correspondence,* 8:47.

210 "About weak points" Ibid., 8:75.

210 "the most perfect of optical" AG, *Darwiniana,* 127.

211 "I see a bird" CD, *Correspondence,* 8:275.

212 "I cannot persuade myself" Ibid.

213 "taking a very great liberty" Ibid., 8:350.

213 "beyond our immediate ken" AG, *Darwiniana,* 129.

214 "a wise man's mind" Ibid., 132.

214 "Most people, and some philosophers" Ibid., 133.

214 "Agreeing that plants" Ibid., 131.

214 "Darwin's particular hypothesis" Ibid., 145.

214 "fortuitous or blind" Ibid., 146–47.

215 "concerns the *order*" Ibid., 149.

215 "may have worn their actual channels" Ibid., 148.

216 "Chance carries no probabilities" Ibid., 153.

Chapter 22: At Down House

219 "Mr Darwin has given the world" Peter Raby, *Alfred Russel Wallace* (London: Chatto & Windus, 2001), 151.

219 "What strikes me" Ibid., 168.

220 "long-continued and intemperate" Horace Greeley, *The American Conflict: A History of the Great Rebellion in the United States of America, 1860–64: Its Causes, Incidents, and Results: Intended to Exhibit Especially Its Moral and Political Phases, with the Drift and Progress of American Opinion Respecting Human Slavery from 1776 to the Close of the War for the Union* (New York: O.D. Case & Co., 1864), 368.

222 "by a single homogenous race" Ibid., 177.

222 **"I hope you have not murdered"** Janet Browne, *Charles Darwin: The Power of Place* (New York: Alfred A. Knopf, 2003), 318.

222 **"The idea of the age"** *LCLB*, 297.

222 **"In attempting to conceive"** CLB, "Darwinism in Germany," *North American Review* 110 (June 1870): 298.

223 **"We (Dr. Gray and I)"** *LCLB*, 303.

223 **"there is no drift toward the worse"** Alfred Emanuel Smith and Francis Walton, *New Outlook* (New York: Outlook, 1875), 525.

224 **"a letter from a clergyman"** *LCLB*, 321.

224 **"I never met a more simple"** Ibid.

225 **"He never stayed long"** Adrian J. Desmond and James Richard Moore, *Darwin* (New York: W. W. Norton, 1994), 562.

225 **"a struggle for existence on our part"** *LAG*, 475.

225 **"if the rebels & scoundrels"** A. Hunter Dupree, *Asa Gray: American Botanist, Friend of Darwin* (Baltimore: Johns Hopkins University Press, 1988), 308.

225 **"Homely, honest, ungainly Lincoln"** Ibid., 309.

226 **"The weak must go to the wall"** *LAG*, 477.

226 **"the South, with its accursed Slavery"** CD, *The Life and Letters of Charles Darwin: Including an Autobiographical Chapter,* ed. Sir Francis Darwin (New York: Appleton, 1898), 2:11.

226 **"by far the best step"** Dupree, *Asa Gray*, 299.

226 **"I had no intention"** CD, *Correspondence,* 8:496.

226 **"I own I cannot see"** Ibid., 8:224.

228 **"the spirit of slavery"** Waldo E. Martin, Jr., *The Mind of Frederick Douglass* (Chapel Hill: University of North Carolina Press, 1894), 247.

228 **"anything like the full force"** Dupree, *Asa Gray,* 280.

228 **"It is foolish to touch"** CD, *Correspondence,* 15:34.

228 **"Can it be reasonably maintained"** Ibid., 15:75.

229 **"there are also mysteries"** Ibid.

230 **"the *Divine* it is which holds"** AG, *Darwinana,* 390.

230 **"We hardly should have thought"** *LAG,* 734.

230 **"so much indebted"** *LCLB,* 443.

Chapter 23: The Ghost of John Brown

231 **"Father said: 'Emerson must see this'"** LMA, *The Journals of Louisa May Alcott,* ed. Joel Myerson, Daniel Shealy, and Madeleine B. Stern (Athens: University of Georgia Press, 1997), 104.

231 **"dusk, could not stop"** Ibid., 103.

231 **"reddest apples and hardest cider"** John Matteson, *Eden's Outcasts: The Story of Louisa May Alcott and Her Father* (New York: W. W. Norton, 2007), 261.

231 **"worked on it as busily"** LMA, *Journals of Louisa Alcott,* 132.

232 **"patriotic blue shirts"** Ibid., 105.

232 **"Here in Concord"** Ellen Tucker Emerson, *The Letters of Ellen Tucker Emerson* (Kent, Ohio: Kent State University Press, 1983), 1:261.

232 "has to choose war stories" LMA, *The Selected Letters of Louisa May Alcott,* ed. Joel Myerson and Daniel Shealy (Athens: University of Georgia Press, 1995), 72.

232 "so riddled with shot" Randall Fuller, *From Battlefields Rising: How the Civil War Transformed American Literature* (New York: Oxford University Press, 2011), 141.

233 "Daresay nothing will ever come of it" LMA, *Journals of Louisa Alcott,* 100.

234 "My dream" Matteson, *Eden's Outcasts,* 349.

235 "This would-be Seer" Rebecca Harding Davis, *Bits of Gossip* (Boston: Houghton Mifflin, 1904), 33–34.

235 "Horrid war" Matteson, *Eden's Outcasts,* 282.

235 "as a person of surpassing sense" Oswald Garrison Villard, *John Brown: 1800– 1859: A Biography after Fifty Years* (Boston: Houghton Mifflin, 1910), 398.

236 "which our colleges fail to deliver" ABA, *The Journals of Bronson Alcott,* ed. Odell Shepard (New York: Little, Brown, 1938), 408.

237 "likeness to the Godhead" Kenneth Cameron, *Concord Harvest Publications of the Concord School of Philosophy and Literature with Notes on Its Successors and Other Resources for Research in Emerson, Thoreau, Alcott and the Later Transcendentalists* (Hartford, Conn.: Transcendental Books, 1970), 79.

237 "Any faith declaring a divorce" Matteson, *Eden's Outcasts,* 374.

237 "Why discuss the Unknowable" Ibid., 392.

239 "the Negro would disappear" George M. Fredrickson, *The Black Image in the White Mind: The Debate on Afro-American Character and Destiny, 1817–1914* (New York: Harper & Row, 1971), 236–37.

240 "I knew John Brown well" FBS, *The Life and Letters of John Brown: Liberator of Kansas, and Martyr of Virginia* (Boston: Roberts Brothers, 1891), 627.

240 "From the first I honored him" Edward J. Renehan, Jr., *Secret Six: The True Tale of the Men Who Conspired with John Brown* (Columbia: University of South Carolina Press, 1997), 269.

Chapter 24: In the Transcendental Graveyard

242 "if Christ did not foresee" HDT, *Journal,* 14:291–92.

243 "fasten myself like a bloodsucker" *HDT,* 155.

243 "cast my vote for Lincoln" ABA, *Diary* (1860), 362.

243 "Confined to my room" Ibid., 368.

243 "No one but a botanist" HDT, *Journal,* 14:3.

244 "the unmeasured and eloquent one" Ibid., 14:117.

244 "I have no doubt" Ibid., 14:146.

245 "We are rag dolls" Loren Eiseley, *The Unexpected Universe* (New York: Harcourt Brace, 1996), 76.

245 "stumps of the oaks" HDT, *Journal,* 14:152.

245 "The pines are the light infantry" Ibid., 14:150.

246 "drive the cattle out" Ibid., 14:151.

246 "the struggle between trees" Alfred Russel Wallace, *Darwinism: An Exposition of the Theory of Natural Selection, with Some of Its Applications* (New York: Macmillan, 1889), 20–21.

246 "constant *new* creation" HDT, *Journal,* 14:147.

247 "a vulgar prejudice" Ibid., 14:311–12.

247 "I took a severe cold" Ibid., 14:290n1.

247 "the union now subsisting" *Journal of the Congress of the Confederate States of America, 1871–1865* (Washington, D.C.: Government Printing Office, 1904), 1:7.

248 "a serious thing" ABA, *The Journals of Bronson Alcott,* ed. Odell Shepard (New York: Little, Brown, 1938), 333.

248 "I *suppose* that I have not many months" Walter Roy Harding, *The Days of Henry Thoreau* (New York: Alfred A. Knopf, 1965), 603.

248 "dying with so much pleasure" Ibid.

248 "Elizabeth Hoar is arranging" Ibid., 604.

249 "This fine morning is sad" Ibid., 605.

249 "lying patiently & cheerfully" Fuller, *From Battlefields Rising,* 34.

249 "The scale on which his studies proceeded" RWE, *The Major Prose of Ralph Waldo Emerson,* ed. Ronald A. Bosco and Joel Myerson (Cambridge, Mass.: Belknap Press, 2015), 467.

249 "A potent presence" LMA, "Thoreau's Flute," *Atlantic Monthly* 12 (September 1863), 281.

250 "Every pine needle expanded" *HDT,* 427.

Selected Bibliography

Adams, Bluford. *E Pluribus Barnum: The Great Showman and the Making of U.S. Popular Culture*. Minneapolis: University of Minnesota Press, 1997.

Adams, Henry. *Democracy, Esther, Mont Saint Michel and Chartres, The Education of Henry Adams*. New York: Library of America, 1983.

Agassiz, Elizabeth Cabot Cary. *Louis Agassiz: His Life and Correspondence*. Boston: Houghton, Mifflin, 1893.

Alcott, Amos Bronson. "Diary," 1860, 1861. Houghton Library, Harvard University.

——. *The Journals of Bronson Alcott*. Edited by Odell Shepard. Boston: Little, Brown, 1938.

——. *The Letters of A. Bronson Alcott*. Edited by Richard L. Herrnstadt. Ames: Iowa State University Press, 1969.

——. *Notes of Conversations, 1848–1875*. Madison, N.J.: Fairleigh Dickinson University Press, 2007.

——. "Orphic Sayings." *Dial* 1 (July 1840).

Alcott, Louisa May. *Hospital Sketches*. Reprint Bedford, Mass.: Applewood Books, 1986.

——. *The Journals of Louisa May Alcott*. Edited by Joel Myerson, Daniel Shealy, and Madeleine B. Stern. Athens: University of Georgia Press, 1997.

——. *Little Women: Or, Meg, Jo, Beth, and Amy*. Boston: Little, Brown, 1922.

——. *Louisa May Alcott on Race, Sex, and Slavery*. Edited by Sarah Elbert. Boston: Northeastern University Press, 1997.

——. *Moods*. Edited by Sarah Elbert. New Brunswick, N.J.: Rutgers University Press, 1991.

——. *The Selected Letters of Louisa May Alcott*. Athens: University of Georgia Press, 1995.

Allen, William G. *American Prejudice Against Color; an Authentic Narrative, Showing How Easily the Nation Got into an Uproar*. London: W. & F. G. Cash, 1853.

Anbinder, Tyler. *Five Points: The 19th-Century New York City Neighborhood That Invented Tap Dance, Stole Elections, and Became the World's Most Notorious Slum*. New York: Free Press, 2010.

Anonymous. "Review, Darwin's Origin of Species." *Dial* 1, no. 3 (March 1860): 196–97.

Anthony, Katharine Susan. *Louisa May Alcott*. Westport, Conn.: Greenwood Press, 1977.

Baker, Carlos. *Emerson Among the Eccentrics: A Group Portrait*. New York: Viking, 1996.

Barton, Cynthia H. *Transcendental Wife: The Life of Abigail May Alcott.* Lanham, Md.: University Press of America, 1996.

Beckert, Sven. *Empire of Cotton: A Global History.* New York: Alfred A. Knopf, 2014.

Beer, Gillian. *Darwin's Plots: Evolutionary Narrative in Darwin, George Eliot and Nineteenth-Century Fiction.* Cambridge: Cambridge University Press, 2009.

Borst, Raymond R. *The Thoreau Log: A Documentary Life of Henry David Thoreau, 1817–1862.* Boston: G. K. Hall, 1992.

Bosco, Ronald A., and Jillmarie Murphy. *Hawthorne in His Own Time: A Biographical Chronicle of His Life.* Iowa City: University of Iowa Press, 2007.

Bowen, Francis. "Transcendentalism." *Christian Examiner and General Review.* Vol. 21. Boston: J. Munroe, 1837.

Bowler, Peter J. *Evolution: The History of an Idea.* Berkeley: University of California Press, 2009.

Brace, Charles Loring. "Diary 1853–55." Children's Aid Society Archive, New-York Historical Society, 55 1853.

———. *Children's Aid Society Seventh Annual Report,* 1860.

———. *The Dangerous Classes of New York and Twenty Years' Work Among Them.* New York: Wynkoop & Hallenbeck, 1872.

———. *The Life of Charles Loring Brace: Chiefly Told in His Own Letters.* Edited by Emma Brace. New York: Charles Scribner's Sons, 1894.

———. *The Races of the Old World.* London: John Murray, 1863.

———. "Winter-Life Among the Poor." *Independent,* February 9, 1860.

Briggs, James A., and Abraham Lincoln. *An Authentic Account of Hon. Abraham Lincoln Being Invited to Give an Address: in Cooper Institute, N.Y., February 27, 1860.* Privately printed, 1915.

Browne, Janet. *Charles Darwin: A Biography.* 2 vols. New York: Alfred A. Knopf, 1995, 2003.

Buell, Lawrence. *The American Transcendentalists: Essential Writings.* New York: Modern Library, 2006.

Burlingame, Michael. *Abraham Lincoln: A Life.* Baltimore: Johns Hopkins University Press, 2013.

Burroughs, John. *The Late Harvest.* Boston: Houghton Mifflin, 1922.

Caldwell, Charles. *Thoughts on the Original Unity of the Human Race.* Cincinnati: J.A. & U.P. James, 1852.

Calore, Paul. *The Causes of the Civil War: The Political, Cultural, Economic and Territorial Disputes between North and South.* Jefferson, N.C.: McFarland, 2008.

Cameron, Kenneth Walter. *Concord Harvest Publications of the Concord School of Philosophy and Literature with Notes on Its Successors And Other Resources for Research in Emerson, Thoreau, Alcott and the Later Transcendentalists.* Hartford, Conn.: Transcendental Books, 1970.

Chambers, Robert. *Vestiges of the Natural History of Creation and Other Evolutionary Writings.* Edited by James A. Secord. Chicago: University of Chicago Press, 1994.

Child, Lydia Maria. *Letters from New-York.* London: Richard Bentley, 1843.

Chopin, Kate. *The Awakening.* Edited by Margaret Culley. New York: W. W. Norton, 1993.

Conway, Moncure Daniel. *Autobiography: Memories and Experiences of Moncure Daniel Conway*. Boston: Houghton Mifflin, 1904.

Dahlstrand, Frederick C. *Amos Bronson Alcott: An Intellectual Biography*. Madison, N.J.: Fairleigh Dickinson University Press, 1982.

Daniels, George H. *Darwinism Comes to America*. New York: Blaisdell, 1968.

Darwin, Charles. *Autobiographies*. Edited by Michael Neve and Sharon Messenger. New York: Penguin Classics, 2002.

———. *Charles Darwin's Beagle Diary*. Edited by R. D. Keynes. Cambridge: Cambridge University Press, 2001.

———. *The Correspondence of Charles Darwin*. Edited by F. Burkhardt et al. New York: Cambridge University Press, 1985.

———. *The Descent of Man, and Selection in Relation to Sex* (Part One), vol. 21 of *The Works of Charles Darwin*. New York: New York University Press, 2009.

———. *Journal of Researches into the Geology and Natural History of the Various Countries Visited by H.M.S. Beagle*. New York: Hafner, 1952.

———. *The Life and Letters of Charles Darwin: Including an Autobiographical Chapter*. Edited by Sir Francis Darwin. New York: Appleton, 1898.

———. *On the Origin of Species: A Facsimile of the First Edition*. Cambridge, Mass.: Harvard University Press, 1964.

———. *The Voyage of the Beagle: Charles Darwin's Journal of Researches*. Edited by Janet Browne and Michael Neve. Abridged ed. New York: Penguin Classics, 1989.

Darwin, Erasmus. *The Temple of Nature, Or, The Origin of Society: A Poem, with Philosophical Notes*. Baltimore: John W. Butler, and Bonsal & Niles, 1804.

Davis, David Brion. *The Problem of Slavery in the Age of Emancipation*. New York: Alfred A. Knopf, 2014.

———. *The Problem of Slavery in Western Culture*. Ithaca, N.Y.: Cornell University Press, 1966.

Davis, Rebecca Harding. *Bits of Gossip*. New York: Constable, 1904.

Dean, Bradley P., ed. "Henry D. Thoreau and Horace Greeley Exchange Letters on the 'Spontaneous Generation of Plants.'" *New England Quarterly* 66, no. 4 (December 1, 1993): 630–38.

Desmond, Adrian J., and James Richard Moore. *Darwin*. New York: W. W. Norton, 1994.

———. *Darwin's Sacred Cause: How a Hatred of Slavery Shaped Darwin's Views on Human Evolution*. Boston: Houghton Mifflin Harcourt, 2009.

Douglass, Frederick. *Autobiographies: Narrative of the Life of Frederick Douglass, an American Slave / My Bondage and My Freedom / Life and Times of Frederick Douglass*. Edited by Henry Louis Gates. New York: Library of America, 1994.

———. *The Claims of the Negro, Ethnologically Considered: An Address, Before the Literary Societies of Western Reserve College, at Commencement, July 12, 1854*. Lee, Mann & Co., Daily American Office, 1854.

———. *Selected Speeches and Writings*. Chicago: Chicago Review Press, 2000.

Dumond, Dwight Lowell. *Southern Editorials on Secession*. Gloucester, Mass.: Peter Smith, 1964.

Dupree, A. Hunter. *Asa Gray: American Botanist, Friend of Darwin*. Baltimore: Johns Hopkins University Press, 1988.

Eiseley, Loren C. *Darwin's Century: Evolution and the Men Who Discovered It*. New York: Doubleday, 1961.

Emerson, Edward Waldo. *The Early Years of the Saturday Club, 1855–1870*. Boston: Houghton Mifflin, 1918.

———. *Henry Thoreau: As Remembered by a Young Friend*. Boston: Houghton Mifflin, 1917.

Emerson, Ellen Tucker. *The Letters of Ellen Tucker Emerson*. Kent, Ohio: Kent State University Press, 1983.

Emerson, Ralph Waldo. *The Correspondence of Emerson and Carlyle*. Edited by Ralph L. Rusk. New York: Columbia University Press, 1964.

———. *Emerson's Antislavery Writings*. Edited by Len Gougeon and Joel Myerson. New Haven, Conn.: Yale University Press, 2002.

———. *Essays and Lectures*. New York: Library of America, 1983.

———. *Journals and Miscellaneous Notebooks of Ralph Waldo Emerson*, vol. 5, *1835–1838*. Cambridge, Mass.: Harvard University Press, 1965.

———. *The Later Lectures of Ralph Waldo Emerson, 1843–1871*, vol. 1, *1843–1854*. Edited by Ronald A. Bosco and Joel Myerson. Athens: University of Georgia Press, 2010.

———. *The Letters of Ralph Waldo Emerson*. Edited by Eleanor Marguerite Tilton. New York: Columbia University Press, 1939.

———. *The Prose of Ralph Waldo Emerson*. Edited by Ronald A. Bosco and Joel Myerson. Cambridge, Mass.: Harvard University Press, 2015.

Fredrickson, George M. *The Black Image in the White Mind: The Debate on Afro-American Character and Destiny, 1817–1914*. New York: Harper & Row, 1971.

Fuller, Randall. *From Battlefields Rising: How the Civil War Transformed American Literature*. New York: Oxford University Press, 2011.

Glick, Thomas F. *The Comparative Reception of Darwinism*. Chicago: University of Chicago Press, 1988.

Goodwin, Joan W. *The Remarkable Mrs. Ripley: The Life of Sarah Alden Bradford Ripley*. Hanover, N.H.: University Press of New England, 1998.

Gopnik, Adam. *Angels and Ages: A Short Book About Darwin, Lincoln, and Modern Life*. New York: Alfred A. Knopf, 2009.

Gould, Stephen Jay. *The Mismeasure of Man*. New York: W. W. Norton, 2006.

Gray, Asa. *Darwiniana: Essays and Reviews Pertaining to Darwinism*. Cambridge: Cambridge University Press, 2009.

———. *Letters of Asa Gray*. Boston: Houghton Mifflin, 1894.

Greeley, Horace. *The American Conflict: A History of the Great Rebellion in the United States of America, 1860–1864: Its Causes, Incidents, and Results: Intended to Exhibit Especially Its Moral and Political Phases, with the Drift and Progress of American Opinion Respecting Human Slavery from 1776 to the Close of the War for the Union*. New York: O.D. Case & Co., 1864.

Gupta, Tania Das. *Race and Racialization: Essential Readings*. Toronto: Canadian Scholars' Press, 2007.

Gura, Philip F. *American Transcendentalism: A History*. New York: Macmillan, 2008.

Harding, Walter Roy. *The Days of Henry Thoreau*. New York: Alfred A. Knopf, 1965.

Hawthorne, Julian. *The Memoirs of Julian Hawthorne*. New York: Macmillan, 1938.

Hawthorne, Nathaniel. *The Marble Faun*. Edited by Christopher Bigsby. London: Everyman, 1995.

Herschel, John F. W. *A Preliminary Discourse on the Study of Natural Philosophy*. Chicago: University of Chicago Press, 1987.

Hickok, Benjamin Blakely. *The Political and Literary Careers of F. B. Sanborn*. Ph.D. diss., Michigan State University, 1953.

Higginson, Thomas Wentworth. *Cheerful Yesterdays*. Boston: Houghton Mifflin, 1898.

———. *Contemporaries*. Boston: Houghton Mifflin, 1899.

———. *Out-Door Papers*. Boston: Lee & Shepard, 1886.

Hochman, Barbara. *"Uncle Tom's Cabin" and the Reading Revolution: Race, Literacy, Childhood, and Fiction, 1851–1911*. Amherst: University of Massachusetts Press, 2011.

Hoeveler, J. David. *The Evolutionists: American Thinkers Confront Charles Darwin, 1860–1920*. Latham, Md.: Rowman & Littlefield, 2007.

Holmes, Oliver Wendell. *The Complete Poetical Works*. Boston: Houghton Mifflin, 1908.

Howells, William Dean. *Literature and Life*, New York: Harper Bros., 1902.

Irmscher, Christoph. *Louis Agassiz: Creator of American Science*. Boston: Houghton Mifflin Harcourt, 2013.

———. *The Poetics of Natural History: From John Bartram to William James*. New Brunswick, N.J.: Rutgers University Press, 1999.

Johnson, Walter. *Soul by Soul: Life Inside the Antebellum Slave Market*. Cambridge, Mass.: Harvard University Press, 1999.

Kaplan, Sidney. *American Studies in Black and White: Selected Essays, 1949–1989*. Amherst: University of Massachusetts Press, 1996.

LaPlante, Eve. *Marmee & Louisa: The Untold Story of Louisa May Alcott and Her Mother*. New York: Free Press, 2012.

Larson, Edward J. *Evolution: The Remarkable History of a Scientific Theory*. New York: Modern Library, 2004.

Lepore, Jill. *The Story of America: Essays on Origins*. Princeton, N.J.: Princeton University Press, 2013.

Levine, George. *Darwin the Writer*. New York: Oxford University Press, 2011.

Lincoln, Abraham. *Complete Works, Comprising His Speeches, State Papers, and Miscellaneous Writings*. Edited by John G. Nicolay and John Hay. New York: Century, 1920.

———. *Speeches of Abraham Lincoln: Including Inaugurals and Proclamations*. New York: A.L. Burt, 1906.

Lindfors, Bernth. *Early African Entertainments Abroad: From the Hottentot Venus to Africa's First Olympians*. Madison: University of Wisconsin Press, 2014.

Longfellow, Henry Wadsworth. *Longfellow's Poetical Works*. London: George Routledge & Sons, 1883.

Lowell, James Russell. *The Complete Poetical Works*. Boston: Houghton, Mifflin, 1896.

Lurie, Edward. *Louis Agassiz: A Life in Science*. Chicago: University of Chicago Press, 1960.

Mackay, Charles. *Life and Liberty in America: Or, Sketches of a Tour in the United States and Canada, in 1857–8*. London: Smith, Elder & Co., 1859.

Marshall, Megan. *The Peabody Sisters: Three Women Who Ignited American Romanticism.* Boston: Houghton Mifflin Harcourt, 2006.

Matteson, John. *Eden's Outcasts: The Story of Louisa May Alcott and Her Father.* New York: W. W. Norton, 2008.

M.B.B. "Darwin's Origin of Species." *Dial* 1, no. 6 (June 1860): 391–92.

McCalman, Iain. *Darwin's Armada: Four Voyages and the Battle for the Theory of Evolution.* New York: W. W. Norton, 2010.

McFarland, Philip. *Hawthorne in Concord.* New York: Grove/Atlantic, 2007.

Mead, Rebecca. *My Life in Middlemarch.* New York: Crown, 2014.

Menand, Louis. *The Metaphysical Club: A Story of Ideas in America.* New York: Macmillan, 2001.

Mitchell, Betty L. "Realities Not Shadows: Franklin Benjamin Sanborn, the Early Years." *Civil War History* 20, no. 2 (1974): 101–17.

Morse, Sidney H., and Joseph B. Marvin. *The Radical.* Boston: A. Williams & Co., 1869.

Myerson, Joel. *A Historical Guide to Ralph Waldo Emerson.* New York: Oxford University Press, 2000.

Norton, Andrews. *A Discourse on the Latest Form of Infidelity: Delivered at the Request of the "Association of the Alumni of the Cambridge Theological School," on the 19th of July, 1839.* Boston: J. Owen, 1839.

Nott, J. C., and George R. Gliddon. *Types of Mankind Or Ethnological Researches, Based Upon the Ancient Monuments, Paintings, Sculptures, and Crania of Races, and Upon Their Natural, Geographical, Philological, and Biblical History.* Philadelphia: Lippincott, 1855.

Numbers, Ronald L. *Darwinism Comes to America.* Cambridge, Mass.: Harvard University Press, 1998.

O'Connor, Stephen. *Orphan Trains: The Story of Charles Loring Brace and the Children He Saved and Failed.* Boston: Houghton Mifflin Harcourt, 2001.

Olmsted, Frederick Law. *The Cotton Kingdom: A Traveller's Observations on Cotton and Slavery in the American Slave States, 1853–1861.* Edited by Arthur M. Schlesinger. New York: Da Capo Press, 1996.

Parfait, Claire. *The Publishing History of "Uncle Tom's Cabin," 1852–2002.* Burlington, Vt.: Ashgate, 2013.

Petrulionis, Sandra Harbert. *To Set This World Right: The Antislavery Movement in Thoreau's Concord.* Ithaca, N.Y.: Cornell University Press, 2006.

Potter, David M. *The Impending Crisis, 1848–1861.* New York: HarperPerennial, 2011.

Prichard, James Cowles. *Researches into the Physical History of Man.* London: J. & A. Arch, 1813.

Raby, Peter. *Alfred Russel Wallace.* London: Chatto & Windus, 2001.

Rawson, Michael. *Eden on the Charles: The Making of Boston.* Cambridge, Mass.: Harvard University Press, 2011.

Redpath, James. *Echoes of Harper's Ferry.* Boston: Thayer & Eldridge, 1860.

Reel, Monte. *Between Man and Beast: An Unlikely Explorer and the African Adventure That Took the Victorian World by Storm.* New York: Anchor, 2013.

Reisen, Harriet. *Louisa May Alcott: The Woman Behind "Little Women."* New York: Macmillan, 2010.

Renehan, Edward J., Jr. *Secret Six: The True Tale of the Men Who Conspired with John Brown.* Columbia: University of South Carolina Press, 1997.

Reynolds, David S. *John Brown, Abolitionist: The Man Who Killed Slavery, Sparked the Civil War, and Seeded Civil Rights.* New York: Alfred A. Knopf, 2009.

Rogers, William Barton. *Life and Letters of William Barton Rogers, Edited by His Wife.* 2 vols. Boston: Houghton Mifflin, 1896.

Ruse, Michael, and Robert J. Richards. *The Cambridge Companion to the "Origin of Species."* Cambridge: Cambridge University Press, 2009.

Russett, Cynthia Eagle. *Darwin in America: The Intellectual Response, 1865–1912.* New York: W. H. Freeman, 1976.

Sacks, Kenneth. *Understanding Emerson: "The American Scholar" and His Struggle for Self-Reliance.* Princeton, N.J.: Princeton University Press, 2003.

Sanborn, Franklin B. "Comment by a Radical Abolitionist." *Century Magazine* 26 (July 1883): 411–15.

——. *The Life and Letters of John Brown: Liberator of Kansas, and Martyr of Virginia.* Boston: Roberts Brothers, 1891.

——. *The Personality of Thoreau.* Boston: C. E. Goodspeed, 1901.

——. *Recollections of Seventy Years.* Boston: R. G. Badger, 1909.

——. *Table Talk: A Transcendentalist's Opinions on American Life, Literature, Art, and People from the Mid-Nineteenth Century through the First Decade of the Twentieth.* Edited by Kenneth Cameron. Hartford, Conn.: Transcendental Books, 1981.

——. *Transcendental and Literary New England: Emerson, Thoreau, Alcott, Bryant, Whittier, Lowell, Longfellow, and Others.* Edited by Kenneth Cameron. Hartford, Conn.: Transcendental Books, 1975.

Sanborn, Franklin Benjamin, and William Torrey Harris. *A. Bronson Alcott: His Life and Philosophy.* Boston: Roberts Brothers, 1893.

Sanborn, Franklin Benjamin, and Benjamin Smith Lyman. *Young Reporter of Concord: A Checklist of F. B. Sanborn's Letters to Benjamin Smith Lyman, 1853–1867, with Extracts Emphasizing Life and Literary Events in the World of Emerson, Thoreau and Alcott.* Edited by Kenneth Cameron. Hartford, Conn.: Transcendental Books, 1978.

Sewall, Richard Benson. *The Life of Emily Dickinson.* Cambridge, Mass.: Harvard University Press, 1994.

Shealy, Daniel. *Alcott in Her Own Time: A Biographical Chronicle of Her Life, Drawn from Recollections, Interviews, and Memoirs by Family, Friends, and Associates.* Iowa City: University of Iowa Press, 2005.

Shepard, Odell. *Pedlar's Progress: Bronson Alcott.* New York: Little, Brown, 1937.

Smith, Alfred Emanuel, and Francis Walton. *New Outlook.* New York: Outlook, 1875.

Specq, François, Laura Dassow Walls, and Michel Granger. *Thoreauvian Modernities: Transatlantic Conversations on an American Icon.* Athens: University of Georgia Press, 2013.

Stauffer, John, and Zoe Trodd. *The Tribunal: Responses to John Brown and the Harpers Ferry Raid.* Cambridge, Mass.: Harvard University Press, 2012.

Stearns, Frank Preston. *The Life and Genius of Nathaniel Hawthorne.* Philadelphia: Lippincott, 1906.

Stern, Madeleine B. *Louisa May Alcott: A Biography.* Lebanon, N.H.: University Press of New England, 2000.

Strong, George Templeton. *The Diary.* Edited by Allan Nevins. New York: Macmillan, 1952.

Tauber, Alfred I. *Henry David Thoreau and the Moral Agency of Knowing.* Berkeley: University of California Press, 2001.

Thoreau, Henry David. *Collected Essays and Poems.* Edited by Elizabeth Hall Witherell. New York: Library of America, 2001.

———. *Faith in a Seed: The Dispersion of Seeds and Other Late Natural History Writings.* Edited by Bradley Dean. Washington, D.C.: Island Press, 1993.

———. *Familiar Letters,* vol. 6 of *The Writings of Henry David Thoreau.* Edited by F. B. Sanborn. Enlarged ed. Boston: Houghton Mifflin, 1906.

———. *The Heart of Thoreau's Journals.* Edited by Odell Shepard, New York: Dover, 1961.

———. *The Journal.* Edited by Bradford Torrey and Francis Allen, 1837–46, 1850– November 3, 1861. 14 vols. Boston: Houghton Mifflin, 1906.

———. *The Journal,* vol. 1, *1837–1844.* Edited by Elizabeth Witherell et al. Princeton, N.J.: Princeton University Press, 1981.

———. *A Week on the Concord and Merrimack Rivers / Walden; Or, Life in the Woods / The Maine Woods / Cape Cod.* New York: Library of America, 1985.

———. *Wild Fruits: Thoreau's Rediscovered Last Manuscript.* Edited by Bradley P. Dean. New York: W. W. Norton, 2001.

———. *The Writings of Henry David Thoreau.* Edited by Franklin Benjamin Sanborn. 10 vols. Boston: Houghton Mifflin, 1896–99.

Villard, Oswald Garrison. *John Brown: 1800–1859: A Biography after Fifty Years.* London: Constable, 1910.

Wallace, Alfred Russel. *Darwinism: An Exposition of the Theory of Natural Selection, with Some of Its Applications.* New York: Macmillan, 1889.

———. *Infinite Tropics: An Alfred Russel Wallace Anthology.* Edited by Andrew Berry. London: Verso, 2003.

Walls, Laura Dassow. *Seeing New Worlds: Henry David Thoreau and Nineteenth-Century Natural Science.* Madison: University of Wisconsin Press, 1995.

Wilentz, Sean. *The Best American History Essays on Lincoln.* New York: Macmillan, 2009.

Wright, Chauncey. *Letters of Chauncey Wright; With Some Account of His Life.* Edited by James Bradley Thayer. Boston: John Wilson, 1878.

Index

Index

Index